FOR DUMMIES™

BESTSELLING BOOK SERIES

GIS For Dummies

W9-BNX-665

Sheet

What You Can Do with GIS

- **Find geographic features.** You can search a GIS database to find point, line, area, and surface features by their descriptions or measurements.

- **Measure geographic features.** You can measure lengths, widths, areas, and volumes, and compare sizes from one feature to another.

- **Characterize distributions.** You can group geographic features and define their distributions based on how much space they use, how close they are to each other, and where they are relative to other features.

- **Summarize geographic data.** You can calculate all sorts of statistics on your geographic features from the simplest descriptive statistics (for example, mean, median, and mode) to very complex spatial statistics.

- **Work with networks.** You can find routes based on time, distance, or other factors. You can route buses to reach the maximum number of people and use population density information to locate stores on these routes near many of your customers.

- **Compare map layers.** You can compare the locations of features from one map layer (or theme) to another. Overlaying the layers shows you the relative location of features from one layer to another.

- **Perform surface analysis.** You can work on the many surfaces available in GIS and use mathematical methods (such as interpolation) to find missing values and perform other analyses.

Map Algebra Functions for Grid-Based GIS

Function Type	Where It Operates	What It's Used For
Local	On individual grid cells	To change cell values based on the value of corresponding grid cells on other layers
Focal	On a specifically targeted grid cell	To return a value (such as an average) based on the values of neighboring grid cells
Zonal	On grid cells in specifically identified regions	To calculate values based on analysis of cells in regions that are not necessarily connected
Block	On square blocks of grid cells	To return a value for the identified block (for example, a 4 x 4 block of cells) on an output grid
Global	On the entire grid	To highlight hard-to-find features and spot general trends across the entire grid
Specialty	On specified grid cells	To perform high-end statistical analysis or create models for moving fluids (such as water or toxic wastes)

For Dummies: Bestselling Book Series for Beginners

GIS For Dummies®

Cheat Sheet

Map Characteristics You Need to Remember

Map Characteristic	What It Means
Maps are models — not miniatures	Maps generalize geographic features by using symbols so that all features will fit the specified output size.
Map scale has a huge impact on GIS analysis	Small scale maps cover large areas with little detail, and large scale maps cover small areas with lots of detail.
Maps are a flat model of a spherical Earth	Maps use projections to compensate for the flat versus spherical issue, and each projection has its own type and amount of distortion.
Maps have a reference grid, or coordinate system	The reference grid helps you navigate the map and links the spherical Earth to the map projection.
Maps have a geographic reference framework, or datum.	Datums are based on a model of the Earth called a reference ellipsoid and enable all the various projections in a GIS to work together to give an accurate picture of the Earth.

Types of GIS Output

- **Maps:** Everyone recognizes this most common output from a GIS.
- **Cartograms:** These special maps distort geographic features based on their values rather than their size.
- **Charts:** GIS can produce pie charts, histograms (bar charts), line charts, and even pictures in addition to maps.
- **Directions:** Directions show or tell you how to get from one place to another.
- **Customer lists:** Business GIS applications often produce customer lists, sometimes with printed mailing labels.
- **3D diagrams and movies:** These forms of GIS output help you see the results of your work realistically and dramatically.

For Dummies: Bestselling Book Series for Beginners

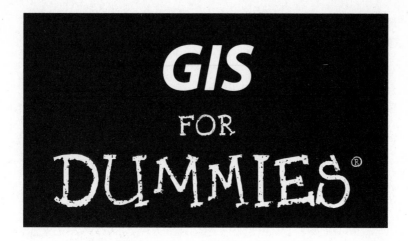

GIS
FOR
DUMMIES®

by Michael N. DeMers

WILEY

Wiley Publishing, Inc.

GIS For Dummies®

Published by
Wiley Publishing, Inc.
111 River Street
Hoboken, NJ 07030-5774

www.wiley.com

Copyright © 2009 by Wiley Publishing, Inc., Indianapolis, Indiana

Published by Wiley Publishing, Inc., Indianapolis, Indiana

Published simultaneously in Canada

For general information on our other products and services, please contact our Customer Care Department within the U.S. at 877-762-2974, outside the U.S. at 317-572-3993, or fax 317-572-4002.

For technical support, please visit www.wiley.com/techsupport.

Wiley also publishes its books in a variety of electronic formats. Some content that appears in print may not be available in electronic books.

Library of Congress Control Number: 2008942757

ISBN: 978-0-470-23682-6

Manufactured in the United States of America

10 9 8 7 6 5 4 3 2 1

WILEY

About the Author

Michael DeMers is an Associate Professor of Geography at New Mexico State University and has been teaching GIS-related courses since 1983. He is the author of *Fundamentals of Geographic Information Systems* (4th Edition), which has been translated into both Russian and simple Chinese, and *GIS Modeling in Raster*, currently being translated into Arabic.

Dedication

To all who have taught me, including my teachers, my students, my colleagues, and those I have known only through their writings.

Acknowledgments

I am grateful to Katie Feltman and Andy Cummings for having the faith in me to write this book. Both gratitude and high praise are due to Colleen Totz Diamond and Laura Miller for their Herculean efforts to make sense of the often vague and always technical GIS ideas and terms. I thank Karen Kemp for her diligent efforts to keep me from technical blunders. My deepest thanks go to Leah Cameron who endured, persevered, and worked tirelessly, all with wonderful humor and constant encouragement. Thanks to all of you in composition, proofreading, page layout, and graphics who convert words and sketches into a polished document.

Thanks to Caliper Corporation, Clarke Laboratories, and Intergraph Corporation for providing me with complementary copies of their software (Maptitude, IDRISI, and GeoMedia Professional, respectively) for the production of this book.

Finally, thanks to all those I love so dearly who have supported me on this journey.

Publisher's Acknowledgments

We're proud of this book; please send us your comments through our online registration form located at `http://dummies.custhelp.com`. For other comments, please contact our Customer Care Department within the U.S. at 877-762-2974, outside the U.S. at 317-572-3993, or fax 317-572-4002.

Some of the people who helped bring this book to market include the following:

Acquisitions and Editorial

Editors: Leah P. Cameron,
Colleen Totz Diamond, Laura Miller

Senior Acquisitions Editor: Katie Feltman

Technical Editor: Karen Kemp

Editorial Assistant: Amanda Foxworth

Sr. Editorial Assistant: Cherie Case

Cartoons: Rich Tennant (`www.the5thwave.com`)

Composition Services

Project Coordinator: Katherine Key

Layout and Graphics: Shawn Frazier,
Nikki Gately, Sarah Philippart,
Christin Swinford, Christine Williams

Proofreader: Debbye Butler

Indexer: Potomac Indexing, LLC

Publishing and Editorial for Technology Dummies

    Richard Swadley, Vice President and Executive Group Publisher

    Andy Cummings, Vice President and Publisher

    Mary Bednarek, Executive Acquisitions Director

    Mary C. Corder, Editorial Director

Publishing for Consumer Dummies

    Diane Graves Steele, Vice President and Publisher

Composition Services

    Gerry Fahey, Vice President of Production Services

    Debbie Stailey, Director of Composition Services

Contents at a Glance

Table of Contents

Introduction

*D*o you plan to purchase a geographic information system (GIS) in the near future? Are you curious about what it can do for you and how you can get the most out of it? Do you need to use the software, or do you need to supervise others who use it? Do you have concerns about how GIS might change the way your organization functions?

If you answered yes to any of these questions, *GIS For Dummies* is the right book for you. GIS is some of the most exciting software to come along in ages, and I want to get you as excited about the possibilities GIS offers as I am. This book can help you start thinking about how you can use maps and harness the awesome power of this new technology.

About This Book

Unlike many books on GIS, this one isn't meant to keep you spellbound for days or weeks. Instead, you can use this book when you need to answer basic questions or figure out what questions to ask your GIS-specialist friends. Think of this book as a reference you can use to find what you need when you need it.

The book gives you a big picture look at GIS — everything from the parts that make up the systems (see Chapter 1) to the spatial information products (see Chapter 20) that the systems produce. So wherever your interests in GIS point you, find those topics in the Table of Contents or Index and jump right in.

Conventions Used in This Book

GIS terminology can get a bit confusing, especially with computer terms. I use the term *raster* to represent both a GIS data structure (composed of square grid cells) and the software based on that structure. When I talk about *vector*, I'm also referring to both the data structure (based on points, lines, and polygons) and software that uses the structure.

When I define a term for you, that term appears in *italics*. Also, I show URLs in `monofont` typeface to set them apart from the regular text.

What You're Not to Read

You may feel the urge, every now and then, to explore some of the more advanced features of the GIS software. The GIS crowd might recognize these features and understand the details, but I don't expect everyone to have (or want to have) that specialized knowledge. Most GIS analysis is based on pretty basic ideas about how things work in geographic space. Sometimes, GIS gets technical and uses fancier methods including mathematical procedures that you probably don't need to understand in depth. So, keep an eye out for the Technical Stuff icons and skip them if you want.

Also, I like to illustrate certain points with extra examples that appear in the book as sidebars. I think you'll find the examples interesting, but they're not essential to your understanding of the basics.

Foolish Assumptions

I'm going to assume that you've heard about GIS but don't know all that much about its inner workings and hidden mechanisms. Many people think GIS (geographic information system) means GPS (global positioning system) because more people have heard the term GPS. In reality, GPS is just a part of GIS, and I tell you about that in Chapter 8. I assume you have something more than a casual interest in GIS, so I explain what GIS is, what it does, and how it can help you with what you do in your organization. Here are a few other assumptions I make:

- **You know what a map is.** GIS relies heavily on maps and map-related data. I assume that you have used a map of some kind, but aren't an expert in either making or using maps. I provide all the background you need to become familiar with how maps represent the real-world geography.

- **You know what geography is.** I assume that you've taken a geography class at some point in your life, but I don't assume that you're a geographer or that you think like a geographer. So I guide you on that path, as well. After you figure out how to think like a geographer (in mapping terms), GIS can become your friend and ally. You might even find it fun to use.

- **You use some form of computer from time to time.** GIS relies on computers. I don't expect that you're a computer technician, but I do assume that you know what data files and software programs are and how to use a computer interface. Beyond that, I explain some of the inner workings of the GIS software and databases so that you can ask intelligent questions of the GIS experts.

How This Book Is Organized

GIS For Dummies contains six parts. They move from general background in geography and mapping in Part I, to the use of computers for maps in Part II, GIS data retrieval in Parts III, pattern identification and analysis in Part IV, a look at GIS output in Part V, and some helpful info about GIS vendors and data sources in the Part of Tens (Part VI).

Part I: GIS: Geography on Steroids

If you're brand new to GIS, you may want to start here. Part I provides a general overview of the book, explains the basic geography background needed to understand how maps represent the real world, and introduces you to some of the mapping terminology that you need to know to understand GIS-speak. It covers map reading, symbolism, *projections* (moving from 3-D to 2-D), *datums* (starting points for measurement), scale issues, and generalization. You can see the power of map data and how getting them into your computer really improves your ability to make use of information contain in maps.

Part II: Geography Goes Digital

Part II deals with how you get data from your paper maps into the computer. If you're unsure about how GIS data work inside the computer, this part can give you the answers. In this part, I show you the two basic models used for digital map representation (grids, called *raster;* and points, lines, and polygons, called *vector*). You find out how these different models enable you to keep track of the geographic features you include in your GIS and how the models link these features to the descriptive information that eventually winds up in your GIS output (such as a map legend).

Part III: Retrieving, Counting, and Characterizing Geography

Part III is for people who want to know how to use GIS to answer questions. It includes information about how to find the geographic features that you put in your GIS database, different ways of searching for features, how to count them up when you find them, and how you can describe what you find. In this part, you discover how to locate and characterize features by type or category, by their sizes and shapes, by measurements that describe them, and even by where they're located relative to each other in geographic space.

Part IV: Analyzing Geographic Patterns

GIS does its most powerful work when analyzing the patterns that you identify, and Part IV focuses on that subject. You see how to measure lengths, areas, distances, and volumes; as well as how to work with networks, such as highways and streets. I explain both topographic and non-topographic surfaces, how to analyze rivers and determine where water will flow during flooding, and how to determine places that an observer can and can't see from a certain point. I even show you how to combine maps and use a powerful map analysis language called *map algebra*.

This part can't make you an expert in GIS analysis, but it can help you figure out enough to start your analyses and talk knowledgeably with the experts.

Part V: GIS Output and Application

In Part V, I show you how to make the most of all your GIS queries and analysis. You can find out about the various types of map output, as well as non-map output, that you can use to help explain the results of your work. I tell you how GIS can generate travel directions, customer lists, alarms, and even movies that show maps through time. Finally, I show you how to smoothly incorporate this high-level technology into your organization so that you can quickly take advantage of its power.

Part VI: The Part of Tens

In the Part of Tens, I introduce ten GIS software vendors and explain what other products and services they provide. I also provide a handy list of questions to ask those vendors before you decide where to purchase software, products, and services. Finally, I also provide a list of sources of GIS data from government and private companies — both free and for purchase.

Icons Used in This Book

GIS For Dummies uses little pictures, or *icons,* that help direct your reading. These little graphics can save you time by letting you find all the high points quickly.

The Tip icon provides a few helpful hints about shortcuts, best practices, and just plain common sense when it comes to GIS. GIS tips help you do the right things at the right time for the right reasons. Each tip comes with an explanation about why it's a good idea, too.

I use the Warning icon to keep you from making mistakes that are very hard to recover from. Unfortunately, GIS doesn't come with many built-in safety mechanisms, so I try to point out potential problem points.

The Remember icon is sort of like a summary of important points that you need to focus on. In some cases, I remind you of things I cover recently in the chapter, and in other cases, I highlight material from other parts of the book and explain how it applies to that specific discussion. Think of them like tiny refresher courses.

The Real World GIS icon highlights all the places that you can find out how people use GIS to accomplish real tasks in the real world.

When you see the Technical Stuff icon, you don't have to read the technical information to understand the surrounding text — but you might want to look at it and get a sense of all the possibilities of GIS.

Where to Go from Here

Because GIS software changes all the time, the user community requires constant updating and retraining. Many fine community colleges, technical schools, colleges, and universities provide formal education in GIS, and some provide continuing education courses to help you keep up with what's going on. Some vendors offer face-to-face and online courses (largely geared toward their product line, of course).

You can also keep updated by interacting with other users. In the past, you'd make these contacts through professional meetings, trade shows, and user group meetings (which still draw plenty of users). Today's technology provides you with a supplemental method of keeping current. Blogs, wikis, forums, and RSS feeds now provide a vast array of methods that you can use to obtain just the right answer in a timely fashion without ever having to leave your computer. Even the vendors themselves often provide audio and video podcasts that give updates on the newest software wrinkles. Here are a few prominent Web-based resources that you can use to continue your GIS education:

- ✔ **GIS Café** (`www.giscafe.com`): A general online GIS community that contains all things GIS, including forums, data providers, and much more.

- ✔ **GIS Data Depot** (`http://data.geocomm.com`): Another general online GIS community which is focused mostly on data sharing.

- ✔ **GIS Lounge** (`http://gislounge.com`): Similar to GIS Café, but with a somewhat more educational and informational spin.

- ✔ **Directions Magazine** (`www.directionsmag.com`): A forum and online GIS magazine that focuses on keeping the reader up to date on research, innovations, software, and hardware related to GIS.

- ✔ **VerySpatial** (`www.veryspatial.com`): A podcast that generally covers the larger discipline of geography, but with a large amount of GIS content.

- ✔ **ESRI Podcasts** (`www.esri.com/news/podcasts/index.html`): ESRI's podcasts include both an instructional series and a speaker series.

GIS forums and podcasts will continue to increase, both in number and in focus. An occasional search on your favorite Internet browser can keep you reading and listening about GIS for a long time to come.

Part I
GIS: Geography on Steroids

The 5th Wave   By Rich Tennant

"According to the GIS generated route, 'Peaceful Haven Overlook' should be just up ahead on the right."

In this part . . .

No, I'm not advocating the use of steroids, but I am advocating GIS. In this part, I provide an introduction to all the geography and map-related capabilities that underlie the enormous power of GIS. You find out how GIS has merged the speed and strength of a computer with a geographer's robust toolkit. And even more impressive, you see how this merger has produced one of the most transforming technologies of the 21st century.

Chapter 1

Seeing the Scope of GIS

In This Chapter

▶ Getting GIS basics

▶ Understanding how all the parts of GIS technology work together

▶ Knowing how to think like a geographer

▶ Seeing the many applications of GIS

*E*verything you experience from day to day happens somewhere in geographic space. As a result, you can represent your world and your experiences in it by using maps. You use those maps to find places, save time while traveling, decide where to locate a new store, plan cities, guide the development of wildlife preserves, and satisfy hundreds of other applications.

In this day of digitized everything, the maps you use to represent the world reside inside the computer, and you now have at your fingertips the ability to search those maps, find objects and routes, and plan related activities. The computer systems that enable you to store and access all this information are collectively called *geographic information systems (GIS).* This book is all about GIS and how it helps you make decisions.

To recognize how important this tool is, you only have to think about how people use maps today and how adding the power of a computer can help you make both better and faster decisions. Here are a few quick examples:

✔ **For business owners and marketers:** You want to put up a new toy store in a location that gives you access to a lot of customers. To find the right location for your store, you need to know who your customers are (parents of children) and where they live (so that you can locate near them). GIS software can find your customers and identify suitable nearby locations where you can locate your store.

✔ **For urban planners:** Say that you're a land-use planner trying to figure out the best way to use a new parcel of land just annexed to the city. To make such decisions, you use mapped information stored in the computer to compare the characteristics of the new parcel to existing land uses and facilities. You can even create a map that shows what the final zoning map should look like.

✔ **For merchandise distributors:** Suppose you work for a wholesale grocer, and you need to move your trucks quickly from store to store by using existing road networks. GIS makes a very complex task — picking the fastest routes for different times of day — very easy for you.

In this chapter, I help you get a feel for how a GIS is organized and how you can use it to make many kinds of decisions — effectively, faster, and with better outcomes than you can with traditional maps alone. Jump in and get ready to enjoy the exciting world of GIS.

Getting a Feel for GIS

A GIS is an operational system that allows resource managers to use some of the tools and skills that geographers use, and a little bit more. Using GIS software, you can put maps and other geographic data into the computer. After you have the data in the computer, you can store, retrieve, and edit that data. You can analyze it (for example, find geographic features, measure distances, or compare patterns) and produce output from it (create new maps from what you find).

Here are some of the things you can do with GIS software:

✔ Selectively retrieve bits and pieces of one or more maps.

✔ Count, group, reclassify, isolate, and quantify features and their patterns on the landscape.

✔ Measure lengths, widths, distances, heights, and volumes of features.

✔ Overlay one map on another to compare features and create new maps.

✔ Visualize, interpolate (predict missing values), slice, cross-section, and generalize surfaces of all kinds.

✔ Track movements and changes in patterns, and predict and exploit pattern change.

✔ Find the shortest, fastest, or most beautiful path, identify potential customers, and locate businesses.

✔ Perform various topographic feature analyses, such as groundwater (subsurface movement of liquid), surface flow (liquids running on the surface), accumulation (liquid gathering in low spots), visibility (places you can see from a specific location), and a host of other analyses.

As the preceding list shows you, GIS is a diverse and powerful tool. In fact, GIS is among the most complex software ever written. It's so complex because it deals with techniques that geographers and related professionals have been devising to analyze maps and map-related data for over 2,500 years. Because of the advances in computer speed and efficiency, new techniques are being added all the time. GIS is truly transformative software: It has the power to change the way decisions are made all over the world. But the software is part of a much larger system, which I describe throughout this book.

Meeting the GIS Collective

The whole of GIS is greater than the sum of its individual parts. And GIS has more parts than just the software and the hardware that drives it (see Figure 1-1). Here are the basic parts that make up the whole GIS:

✔ Data and information

✔ Computers, input and output technology, and computer software

✔ Geographic and related concepts that drive the analysis

✔ People, such as operators, managers, consultants, vendors, and so on

✔ Institutions and organizations within which the GIS exists

Figure 1-1:
GIS is a
collection
of software,
hardware,
data, and
people.

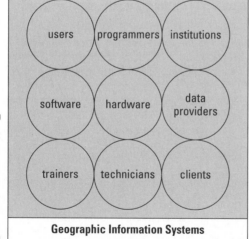

Accumulating geographic data

Projects are the driving force behind GIS analysis and products, but data provide the fuel. Without relevant, timely, accurate data, the GIS can do little to assist decision makers. GIS works with many types of data, but you can group them into two basic categories: primary and secondary. *Primary data* are collected firsthand by you, for a particular project. Primary data are usually the best data for the job because you collect them with your specific goals in mind. *Secondary data* come from others who collect the data for unrelated tasks or gather it with remote sensors.

To get data into a GIS, follow these steps:

1. **Define how and where to best gather, acquire, or sample the data.**

   Defining how and where to collect data involves decisions about whether you want to use existing GIS data or maps, collect field data, perform a census, take polls, interpret aerial photographs, take GPS readings, or use satellites that provide images of the Earth. Deciding what type of data you want to collect requires some specific idea of your data needs and how you can best meet those needs.

2. **Collect, or *sample,* data directly or indirectly by using sensors.**

   The process of collecting depends on your choice of data. The variety of input comes with many possible questions regarding location, scale, sampling schemes, instruments, projections, datums (see Chapter 2), and time. A good collecting methodology, combined with good instrumentation, can have a huge impact on the quality of analysis that your GIS can perform.

3. **Transform the data into a form the computer can recognize.**

   Having good data doesn't help your GIS if you can't get it into the computer. If you're lucky, the data you need are already in a format that your GIS can use. Otherwise, the process of moving analog or non-GIS-compatible data into the computer can be a major part of the GIS operation — both in terms of time and money. You may need to change some data from hard copy to digital forms; you may need to convert some from uncategorized to categorized data (for example, aerial photo interpretation); and you may need to attach coordinates to digital data so that you can find them in your digital maps.

Adding the right computing power

Your computer houses the software that does much of the work of analyzing and manipulating the GIS data. But you need many other forms of hardware and software components working together for analysis to happen.

Hardware related to GIS comes in four flavors:

✔ **Data collection devices:** Collection devices include soil probes, GPS units, analog and digital cameras, voice recorders, satellite remote sensing devices, and telemetry devices. Many of these devices have their own data formats and software that your GIS must be able to work with. Fortunately, most equipment vendors recognize this need, so their output is GIS compatible.

✔ **Data input devices:** Input devices include the computer itself, which you can use to trace lines or collect points from images right on-screen; manual digitizers (basically, really big mouse pads that include position sensors and very sensitive mice); and scanners, both little flatbed scanners and really huge drum scanners that you can use to input big maps. All these devices have their own software and graphics languages that your GIS translates for you.

✔ **Data storage devices and analysis software:** The computer that holds the GIS software also provides storage (both internal and external) and other programs that allow you to analyze lots of map data.

✔ **Data output devices:** Output devices normally include your computer monitor with its many different sizes and number of pixels, printers of all sorts and types, and plotters. Some output devices (such as the monitor) are better for looking things over while you work, some are good for small-sized output (printers), and others do well for larger-sized output (plotters). Some output devices are effective for only small numbers of copies, and others work well when you have large, production-scale jobs. You need to decide which of these devices fits your GIS needs.

The software that comes with input and output devices is usually designed for those specific types of devices. You get the software when you buy the hardware itself. But with GIS, you get to pick out the software that you'll use with your hardware. You can find very simple GIS software and very complex GIS software. Your GIS software might prefer that its data look like little squares *(grid cells),* or it might want more graphic-like data. Some GIS software includes a large suite of programs for many uses, but others might be quite specific to a particular set of tasks (such as working with roads and other networks). You need to know your GIS needs and talk to many vendors to find the GIS software that's right for you.

Computers differ by type and by operating system. Be sure that the software you choose for your GIS work is compatible with your computer. Also, don't buy some GIS software simply because it's exciting or the most complete — or because it's the most expensive. Know your needs and match the hardware and the software to those needs. You always have time to grow bigger and upgrade later, if you need to.

Providing display and representation

GIS is more than a geographic analysis tool; it's also a display and representation tool. In fact, when you look at preliminary results on your screen, those results often give you ideas about what else you might want to do or how you want to proceed with your analysis. Plus, you can now produce the outcome of your work in much more interesting ways, for example, with animations and perspective views that allow you to examine your data more thoroughly.

Here are just some of the ways that people can use GIS in their daily lives:

- ✔ An environmental scientist can drape the vegetation from one map over a 3D view of the topography so that he can visualize the relationships between elevation and vegetation.

- ✔ A Realtor can pull up a map that shows six houses within the client's price range. By clicking each house, the Realtor can also get text information, digital photographs, and even a virtual tour of each house to help the client decide whether to visit.

- ✔ At a city council meeting, an urban planner wants to demonstrate where the city is growing, so she uses a series of maps of the city displayed as a short animated film.

- ✔ A Native American group combines GIS-based maps of their lands with verbal history files compiled as hot-linked digital movies showing the actual residents. The group can then use these maps to help preserve details of their culture for future generations.

- ✔ At a national park, patrons can use a special user interface to locate parts of the park on a map, and the software takes them on a virtual fly-through, showing them the features from a hang glider's perspective.

- ✔ A field agent for a company wants to show his land inventory supervisor an aerial photograph of the company's land, including the parcel lines. He uses his laptop GIS software to send that photograph directly to his hand-held device so that together they can make decisions immediately.

GIS software offers a ton of possibilities, and those possibilities keep expanding while the technology continues to get faster.

Working with people

GIS software is sold by people, to be used by people, to make decisions that affect people. These people may work in business, government, military, education, not-for-profit, medical, and hundreds of other types of organizations. GIS helps these organizations do their work effectively. Because it's high-end technology (requiring a major investment in systems and training), GIS also changes the way organizations work.

Organizations that use GIS work best when the organization adapts itself to the technology. If GIS helps the organization perform its tasks, if the employees are adapting to and benefitting from the changes, if the organization provides training, and if GIS enhances the organization's overall goals, that organization can likely incorporate GIS successfully, long-term. The vast majority of failures to successfully incorporate GIS are a result of poor design or, in most cases, no design at all.

Fortunately, there has been a growth in the number of GIS specialists called *system designers* who help organizations integrate GIS into their practices. A system designer reviews an organization's structure, products, workflow, and needs. He or she then determines the costs and benefits of GIS for that organization, as well as how the organization might best include GIS in critical operations. If you're thinking about GIS, first look for a system design professional. I explain some of the ins and outs of GIS design in Chapter 20.

Knowing How to Think Spatially

All tools are designed to meet a certain need. Sharp things are for cutting, hard things for hammering, and pointy things for binding things together. GIS is also a tool. Geographers, who needed a tool (more like a toolkit) to answer geographic questions, were the primary designers of the original GIS software. Today, many disciplines have contributed enormously to this growing field, but the questions GIS is designed to answer remain fundamentally geographic ones.

To get the most out of your GIS experience, you need to get familiar with some geography. By geography, I'm not talking about knowing the capital of Lichtenstein or the third largest selling product in Wisconsin. Knowing those facts might come in handy when you play Trivial Pursuit, but that information can't help you think like a geographer.

Thinking like a geographer means that you see the world primarily as maps. You see maps everywhere, and those maps show you patterns, distributions, and co-occurrences of geographic features. Think of how your occupation, your business, or your recreation depends on location and distance. The maps you examine begin to come alive with questions such as these:

- Can I sell a lot of clothes if I put my new clothing store next to the new subdivision?
- What's the easiest way to get downtown during rush hour traffic?
- Where's the best place for me to build my vacation home?
- Does the north or east end of town have more gang activity during the summer months?
- Why does such a high incidence of cancer mortality occur in this part of the state?
- What reasons might account for the gap in the distribution of this bird species?
- Where will traffic congestion be most problematic in ten years, given the current patterns of population change?

Geographers answer these kinds of questions, and they use GIS to help figure out the answers. GIS allows them to identify, characterize, question, analyze, explain, and finally exploit their knowledge of patterns and distributions. You don't need to be a geographer to think geographically, but you do need to think geographically to take full advantage of GIS.

Recognizing the spatial nature of questions

Geographers know that all things are related in geographic space, but close things are more related than far things. This statement describes one aspect of geographic space — closeness — that makes space so important to you as a geographic decision-maker. For example, your business might be more successful if you could position it closer to your customers. And the closer you are to the beach, the more likely you are to include swimming in your daily exercise routine.

You encounter many other aspects of space in everyday life. Each aspect of space has an effect on what you do. Here are some aspects of geographic space to help get you thinking spatially:

- **Density:** If you are an urban planner, the more houses in an area (the higher the density), the more potential riders you have for a public transit system.

- **Sinuosity:** Ever notice how those winding subdivisions force you to drive slowly? Urban planners design the winding streets purposefully so that pedestrians don't have to worry about getting hit by high-speed traffic.

- **Connectivity:** People find it difficult to get to some small towns in isolated parts of countries, which shows the impact that a lack of connectivity can have on your ability to get from one place to another.

- **Pattern change:** Over time, a lot of agricultural landscape becomes very fragmented by pockets of residential neighborhoods. What impact might this change have on future agricultural production or habitat for deer?

- **Movement:** Weathermen track the paths of hurricanes, and when they can predict where the storms will travel next, they can save lives.

- **Shape:** A developer wants to buy a piece of land, but wants to make sure that it's square, not oblong, so that the new big house to be built will have plenty of space on all sides.

- **Size:** As a farmer, you need large chunks of land because you're using large equipment that doesn't work in small parcels.

✔ **Isolation:** You're a business owner, and you notice that the number of people who visit your store is dropping. Do you suppose this decline is happening because all the surrounding stores have gone out of business?

✔ **Adjacency:** The value of your home goes down because the zoning board just rezoned the adjacent parcel for commercial use.

All these factors and many more have one thing in common — they all require an ability to see, acknowledge, and question spatial locations, patterns, and distributions. Many fields have certain patterns that they use. Business owners watch the bottom line, historians watch events through time, and meteorologists watch the changes in weather elements. To work effectively in their fields, these people must gain a heightened awareness of the features and patterns relevant to their own occupations. Likewise, GIS practitioners have to make sense of the patterns in spatial information — so they can use that information to make good decisions.

So, how do you become more sensitive to spatial patterns? Like anything else, it takes practice. If you want to see patterns in maps, look at a lot of maps. You come to recognize patterns of traffic by consciously thinking about the traffic density and flow when you drive. Aerial photographs give you a bird's-eye view of land patterns. Satellite images show major patterns that cover large chunks of the Earth.

Read maps, study aerial photographs, peruse satellite images, and — most important — practice creating, querying, and analyzing spatial data with GIS. The more you work with it, and the more you interact with other users, the more easily you can see patterns. You can soon be on your way to thinking in GIS terms — spatially.

Discovering what's so special about spatial data

Thinking spatially helps you work with GIS data after you put those data in the computer. But to become really proficient, you have to know something about how to structure geographic data so that the computer can think more like you do. Because computers don't read maps, you have to exactly identify each item, including its name, its values and properties, and how many dimensions it has. But if you define every type of object that makes up your data correctly, your GIS software knows what to do with it. The following sections cover the two most important aspects of geographic data that your GIS needs to know: dimensionality and descriptions.

Understanding points, lines, polygons, and surfaces

Geographic data come in four basic forms: points, lines, polygons (or areas), and surfaces. A fifth form, related to surfaces, is volumes. Each of these types of data brings with it a particular dimensionality and an associated set of descriptive characteristics. You determine the dimensionality and characteristics of geographic data by how you conceive them, at what scale you observe them, and for what purpose you intend to use them.

Here's a quick summary of these different dimensions and characteristics:

- ✔ **Points:** Zero-dimensional features. Because every object takes up space, no geographic feature can really have *no* length or width. Instead, you can use points to represent geographic features on maps. You can think of a city as a point if you simply want to examine its name and location on a large regional map, for example. Obviously, a city has length and width, but at a certain scale, you can represent it as only a point. (Chapter 2 tells you all about map scale and its effect on geographic features.) Points on your map and in your GIS might include objects such as wells, churches, houses, towns, or trees, depending on the scale you're working with.

- ✔ **Lines:** One-dimensional features, or features that have length. Many line features also have width in real life, but like with points, GIS software assumes that they don't. Features that a cartographer depicts as lines include roads, rail lines, fault zones, fences, and hedgerows. Lines that generally do have no width in real life are those that indicate the separation between two or more places, such as state, local, or national borders. Because you know that all lines inside the computer have length, you can measure characteristics such as length and sinuosity in GIS.

- ✔ **Polygons:** More commonly known as *areas*. The two dimensions of areas — length and width — give you some serious power because you can use them to measure perimeter, area, shape, and the longest and shortest axis by using your GIS. Typical areas that you can find in a GIS include political regions (such as states and countries), land-use and land-cover zones, fields, and many more.

- ✔ **Surfaces:** Have three dimensions — length, width, and a third dimension determined by the characteristics of the surface. Physical surfaces allow you to calculate volumes, predict trends, and determine the direction of the flow of fluids. You may work with many types of physical surfaces, as well as some nonphysical surfaces. You might encounter topographic surfaces, air pressure surfaces, bathymetric (underwater) surfaces, economic surfaces, and population surfaces.

You can examine each of these types of geographic objects individually. The fun starts when you compare them and work with them all at the same time. You can give these objects names and a lot of other descriptions that you may want to use in your analyses.

Putting names and numbers on spatial data

GIS is more than graphics. Most systems contain database tables that allow you to store all sorts of descriptive information about the points, lines, areas, and surfaces that you're depicting in your GIS. The nature of database tables requires you to be just as picky about assigning descriptive information to your objects as you are about choosing the right graphics to depict the objects themselves. You need to understand the basics of consistent data description because many GIS mistakes (for example, retrieving the wrong data) can occur if you incorrectly mix descriptions when you do data analysis.

Here are the levels of description and a brief explanation of each:

- ✔ **Nominal:** Geographic features that have names only. So, you can't compare their descriptive information to any other. Examples of nominal descriptions you might find on a map include the *Bethel Synagogue, Lake Drive, Maple Township,* and *Mount Flattop.*

- ✔ **Ordinal:** Geographic features that you can compare by rank. You could have short, medium, and tall trees; dirt roads, paved roads, highways, and superhighways; or large, medium, and small chemical spills.

- ✔ **Interval:** Geographic features that have detailed increments *(intervals)* that you can measure. One limiting characteristic of interval data is that, although you can get very accurate measurements, you can't form ratios because the starting point is arbitrary. For example, if the soil in land parcel A is 15 degrees centigrade and the soil in parcel B is 30 degrees, you can say that soil B is 15 degrees warmer than soil A, but you can't say that it's twice as warm because 0 degrees centigrade is an arbitrary starting place and the temperature values can thus be negative.

- ✔ **Ratio:** Geographic data that have measurable units, like interval data, but also allow you to make the ratio comparisons that interval data won't. If you own a parcel of land that's worth $20,000 and your neighbor has a parcel worth $10,000, then your parcel isn't just worth $10,000 more, it's also twice as expensive. The key point is that ratio data have an absolute 0. A person can't have a negative weight, and the number of people can't be a negative value.

- ✔ **Scalar:** Scalar data are a bit difficult to define, but here's my take: Scalar data have a proprietary measurement system. That is, you create your own scale that applies to only a particular set of data. So, if you're ranking the beauty of a scenic overlook on a scale of 1 to 10, you first have to decide what each number in the scale means. GIS allows you to establish a scalar description for features that you can't really measure any other way.

The computer represents even nominal data (names) with numbers in the GIS database. Try to avoid using mathematical techniques that force you to multiply the numbers that represent nominal categories by ordinal, interval, or ratio numbers. Attempting to multiply a nominal category *(urban)* by a ratio category *(feet)* often yields downright silly results (in this case, *urban feet*).

At Least 101 Uses of GIS

GIS is like duct tape; it has many, many uses. You can find modern professional GIS software for all types of users. Because you're reading this book right now, you probably already suspect (or *know*) that GIS can help you. Some applications (such as managing land records or natural resource inventories) are pretty standard and provide valuable, timely information. Other applications (such as predicting flood potential or estimating potential hurricane damage) are more experimental and even stretch the capabilities of the GIS.

Managing business activities

GIS is about business. In fact, a whole new industry of business geographics is developing because GIS can affect the bottom line, improve product quality, and provide new opportunities. Here are a few examples of how different sectors of the business world are using GIS:

- **Banking:** GIS activities in banking and financial institutions include regulatory compliance (such as home mortgage disclosure act regulations), customer prospecting, and locating new branches and ATMs.

- **Business locations and customer behavior:** You can more effectively figure out where to place a business if you know your customer base's location. The closer or more accessible you are to the people whose buying patterns match your products, the more apt you are to be successful. Businesses can also compare market share with other surrounding businesses and adjust to changing demographic conditions.

- **Insurance claims adjustment:** GIS can perform all the typical business and marketing tasks, including identifying potential clients, and determining risk factors. And if you're an insurance broker, imagine being able to match applications for storm and flood insurance with a map of the 100-year flood zone.

- **Journalism:** The media, especially the television news media, use flythroughs, zoom-ins, and visual overlays of maps on imagery, providing a media-rich and enticing product.

- **Real estate:** Real estate agents, both residential and commercial, use a GIS (the multiple listing service) to search for and select properties that match client needs. Some appraisers use GIS to perform mass appraisals of whole regions at the same time.

- **Trucking and delivery:** Moving products and material is getting more expensive all the time. Minimizing route lengths reduces cost, speeds delivery times, and increases customer satisfaction. GIS has tools specifically designed to work with road and rail networks.

Planning city operations and expansion

City, county, and regional planning has long used GIS to track development, zone land parcels, assess available resources, and plan for future growth. GIS allows planners to evaluate master plans, monitor expansion and traffic patterns, predict change, monitor population, and even decide the best place to put the new government planning office. Many GIS operations across the world are called planning departments, not GIS departments. But they often use GIS as their primary tool.

Providing protection and emergency services

Police, human services, and emergency services (the Enhanced-911 call-routing service, for example) are beginning to use GIS. Crime mappers can identify crime hot spots and move officers where needed, corrections officers can track their parolees, hospitals can be placed where they meet the most need, and dispatchers can route emergency services (such as ambulance and fire) to their destinations — all with the power of GIS.

Land management and conservation

The first major implementation of GIS managed the enormous expanse of Canada's natural and mineral resources. In this and similar applications, the software can help monitor fires and dispatch firefighters, monitor and manage disease and insect outbreaks, control land use and land inventory, select set-asides and easements, track and manage wildlife, plan for ecotourism, and much more. Combined with today's expanding complement of Earth-sensing satellites, the role of GIS in land management is sure to continue to expand.

Military and defense-related tasks

The military and intelligence communities are taking advantage of the GIS toolkit, which includes tools specifically targeted to those users. By combining top-secret satellite data and visual evaluation from unmanned aerial vehicles (UAVs) with the power of GIS and existing datasets, defense departments can evaluate troop movements, target artillery fire, test scenarios, perform supply and logistics operations, and monitor borders. The military and intelligence communities often rely on the same geospatial tools available to the general public — but they have exclusive access to certain data and data sources, as well as some additional and quite sophisticated software.

A treasure chest of possibilities

Really, the potential uses of GIS have no limit. Jack Dangermond, the president of Environmental Systems Research Institute (ESRI), has been quoted as saying, "The application of GIS is only limited by the mind of the user." I've seen GIS used to reconstruct American Revolutionary War battles, follow the Salem witch trials, track debris from falling space equipment, assist hotline psychics (yes, you read that correctly) with profiling their callers, monitor honey bee movement, control farm equipment, tell ranchers where to move their livestock, and plan interesting entertainment sites.

While traditional applications of GIS technology remain the mainstay of the industry, GIS uses continue to expand. GIS capabilities also continue to increase while humanity's knowledge of geography grows and people encode that knowledge into databases and GIS algorithms. If you're getting into GIS now, you'll probably spend much of your time figuring out both the concepts that drive GIS and the new software that keeps coming out — just when you think you know what to do with your existing software.

If you enter the term `geographic information systems` in an Internet search engine, such as Google (`www.google.com`), you retrieve well over 11,000,000 hits. You may find the volume of information a bit overwhelming, but you can keep up with all the many changes by looking to GIS faculty at colleges and universities, GIS company Web sites, and online GIS communities.

You get over 50,000,000 hits if you enter the keyword GIS in an Internet search engine, but many of the results you get deal with U.S. military GIs, not geographic information systems.

Chapter 2

Recognizing How Maps Show Information

Maps are a compact and elegant method of communicating information. Most folks have used maps in one form or another for a variety of purposes. You've probably used road maps to help you find your way from one town to another and street maps to get around in a city or town. The maps in newspapers show what's happening in different parts of the world. An atlas may be just the key you need to figure out the name of a mountain for a crossword puzzle. If a map is designed correctly, you can find much of what you're seeking without any help at all, and you can interpret the data with little effort. This chapter sets you on course to understanding the sometimes complex, but usually intuitive, language of maps.

Knowing How Maps Represent Geography

Like with any form of communication, you must apply some basic standards before a map can effortlessly communicate the information it contains. Even when you sketch a simple map on a piece of paper, the first decision you must make is how much space you need to create the map. You can think of this area as your map *extent*. When you use a GIS to either create a map or analyze map data, you make decisions about how much area you want to work with. The area of your digital map is related to the actual amount of geographic space that you're examining, but it's much smaller than the real geographic space. This relationship, called *map scale,* also determines how big you can make the graphic objects (symbols) and how many of those objects you can put on the map.

Understanding scale

Naturally, the portion of the Earth represented on a map is much larger than the drawing. The map and the Earth are at different *scales.* Scale determines how much or how little detail your map can hold.

You can often find a map's scale represented by a graphic bar and a fraction that shows the relationship between the size of the map in the *numerator* (the fraction's top part) and the size of the Earth in the *denominator* (the fraction's bottom part). Using this mathematical approach, the smaller the fraction (one with a small numerator and a large denominator), the smaller the scale. The smaller the map scale, the larger the amount of the Earth that map represents.

To get this representation to work, the mapmaker *(cartographer)* uses a little trick or two. Instead of including all the detail of the Earth, he or she selects which objects to show and decides how much graphic detail of each object he or she can omit to save space while still expressing the necessary information. Table 2-1 — from the U.S. Geological Survey of typical map scales — shows the amount of the Earth that different map scales represent.

Table 2-1	Map Scales	
Scale	*1 inch represents approximately*	*1 cm represents*
1/20,000	1,667 ft.	200 m
1/24,000	2,000 ft. (exact)	240 m
1/25,000	2,083 ft.	250 m
1/50,000	4,166 ft.	500 m
1/62,500	1 mile	625 m
1/63,360	1 mile (exact)	633.6 m
1/100,000	1.6 miles	1 km
1/125,000	2 miles	1.25 km
1/250,000	4 miles	2.5 km
1/500,000	8 miles	5 km
1/1,000,000	16 miles	10 km
1/2,000,000	32 miles	20 km

Both the mapmaker and the map user need to know map scale so they can (respectively) create and read the map properly. I show how map scale affects your GIS operations in the section "Grasping the importance of scale," later in this chapter.

Interpreting symbols

Shades, lines, shapes, and words are the language of a map. A map may contain various shaded lines, little white squares, circles, gray patches, words, letters, and numbers. You may find the number of symbols a little overwhelming. As a GIS and map user, you really need only to recognize that all the symbols do have a purpose on the paper. They don't just happen (much to the chagrin of the people who have to make them). Cartographers plan out the selection, size, placement, and design of all these symbols before putting those symbols on the map. Figure 2-1 shows a road map, representing part of England, that contains various symbols.

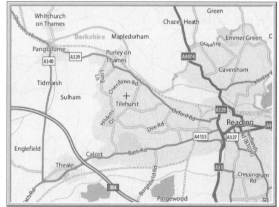

Figure 2-1:
Symbols on
a map of
England.

At first glance, the map in Figure 2-1 may appear disorganized. But each symbol appears in a particular place because it represents some real-world feature that exists in more or less the same place on the Earth. For example, the city of Reading is on the right side of the map, and Pangbourne is to the left and up a bit. This positioning tells you that Pangbourne is north and west of Reading (if you assume that north is toward the top of the map). Because you can relate the symbols to the features they represent, you can quickly orient yourself to the map. Without ever visiting England, you can see how the various towns are arranged in space relative to one another. You can also see that lines, which represent roads, connect these point symbols.

Chapter 3 explains how to translate map symbols.

Incorporating symbols into your map

A map's symbols show you what geographic features are represented on the map and where they occur. Cartographers use symbols that represent point features (such as towns), symbols that represent linear features (such as roads and rivers), and symbols that represent area features (such as lakes and towns). I know, I know — I said towns twice! How can towns be both points and areas?

The short answer is that choosing the right symbol to represent a particular feature depends on a map's scale. If you live in a big city, you might have a whole fold-out map to represent the area that your city occupies. That map is at a fairly large scale (perhaps 1:30,000) and is designed to show small areas in detail. If you look at your city on a map in a world atlas, however, that city may show up as a single dot. Figure 2-2 shows how the same geographic feature looks on maps of different scales.

Figure 2-2:
A city shown on large-scale and small-scale (upper-right inset) maps.

The cartographer has to decide when to represent area features as points and when to represent them as areas. One of the biggest headaches that cartographers deal with is deciding how to generalize map symbols to represent reality as best they can. The factors that they have to consider when they make such decisions include the following:

✔ **Scale:** Determines how many geographic features can be symbolized on a map. When cartographers have tons of data to represent, their symbol choices have to take into account map readability. So, they make decisions to try to avoid symbol crowding. The cartographer has to decide which features to include, which to omit, which to simplify, and which to move to make room for other symbols.

✔ **Data availability:** Determines what type of information can be put on the map. Cartographers can't make up data (if they want their maps to be accurate, that is). For example, suppose that the cartographer wants to portray 12 land-use types on a 2008 land-use map, but has data for only 5 land-use types from 1968. In this case, he or she won't be able to produce the 2008 land-use map without collecting more data.

✔ **Limitations of output devices:** The cartographer also has to consider how symbols will print. He or she must know which features are most important to show (at the map scale to be used for printing) and plan the symbols to use accordingly.

✔ **Reader characteristics:** Not all readers have 20/20 vision, color vision, or any vision at all. Moreover, some maps are designed to be read close-up (a few inches from your face) while others are displayed on walls and read from a distance. So, who reads the map and how they read it affect decisions about

   • Whether to use color or black and white

   • How large or small to make the scale, symbols, and fonts

   • Whether to include tactile symbols for vision-impaired map users

Many traditional maps serve as input to your GIS, so all the generalizations that cartographers make impact the quality of the analysis that results from the system. Also, when you're ready to produce your own maps (Chapter 19), you need to be aware of these issues.

Recognizing the Different Types of Maps

Anyone who creates a map must consider the content and subject of the map. He or she must assign a topic, or *theme,* to the map and set the individual details in the proper context. If you're creating a map to your garden party, for example, you must include roads so that your guests know what pathways to travel. Landmarks and buildings may also help the reader get oriented to your part of town. You may also include text on the map, such as, "Turn left at the ugly purple house."

In a technical conversation, words often have different or more precise meanings than in a general conversation. The context and the theme are just as important in mapping as in a general conversation. A map needs a unique context, a specific vocabulary, and a set of rules that pull all these parts together so that both the mapmaker and the map reader can understand each other and make the best possible GIS.

The easiest way to categorize maps is to separate them into two basic groups — reference maps and thematic maps, as described in the following sections.

Reading reference maps

Reference maps offer a great deal of information on a single document. Atlases generally contain reference maps so that many related maps can be contained in the same place. Reference maps often cover very large portions of the Earth, which means they're created at small scales. Therefore, reference maps aren't designed to provide extremely accurate depictions of locations. Instead, reference maps offer only the pattern and general locations of things.

Reference maps also commonly contain several types of information on the same map. In this way, the map helps orient users to the geography of the area represented, as a whole. A typical reference map contains features such as regional borders; names of major cities; major transportation routes; and often prominent physical features, such as major rivers, lakes, and mountain ranges. Figure 2-3 shows a reference map of the Middle East from the CIA World Fact Book, which includes regional borders, transportation routes, and physical features.

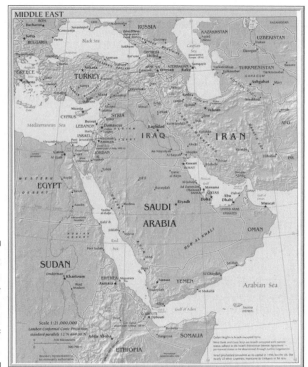

Figure 2-3:
Reference maps may contain several types of information.

Using thematic maps

A thematic map provides as much accurate, detailed information as possible about a particular subject, such as roads or hills, as compared to a reference map (discussed in the preceding section), which tries to select the most important information about several subjects.

When you view a reference map, such as the one shown in Figure 2-3, you can see a number of themes of data — in this case, regional, transportation, and physical features. The reference map doesn't present these themes in much detail, and it doesn't have the primary purpose of communicating specifics. For example, although the map includes transportation information, that information isn't complete — it's very general because of the small scale of the map, and the different types of routes aren't clearly identified. This map wouldn't help you much if you were trying to navigate within Saudi Arabia or Iran.

For navigating, you need a *road map,* as shown in Figure 2-1. A road map is one example of a thematic map because it focuses on communicating information about roads. Although such a thematic map might include a few regional subdivisions and other landmarks, the majority of the map is designed so that you know which roads are two lane and which are one lane, which are dirt roads and which are superhighways, which are one-way streets and which are two-way streets, and so on.

Thematic maps are the primary kind of maps that you use in your GIS activities. Although you might find reference maps useful for some projects that involve large study areas, the small scale of reference maps usually limits their usefulness for analysis. The more familiar you become with the wide variety of thematic maps available to you, the more easily you can recognize which maps might assist you in your GIS work.

The list of potential thematic maps is staggering; here are a few examples:

- **Medical maps:** Cancer mortality, spread of epidemics, and health hazards can be map subjects. Medical professionals may use such maps to determine characteristics of locations where some diseases occur, concentrate, and spread. Knowing these factors can help them in their goals to prevent or cure disease.

- **Hazardous material maps:** These maps give transportation companies information about where trucks carrying hazardous material can travel, as well as the relative locations of people at risk (for example, the elderly or hospital patients).

- **Law enforcement maps:** These kinds of maps show the police crime hot spots, locations of officers, crimes by time of day and season, and socio-economic factors that might affect crime rates.

✔ **Utility maps:** Gas and electric utility infrastructure maps can include data about the type of equipment, its condition and age, and many other factors affecting the energy companies' ability to provide and sustain services.

If a feature appears or an event occurs on the surface of the Earth, it can be — and often has been — mapped.

Grasping the importance of scale

When compared to thematic maps, reference maps are often created on too small a scale to be particularly useful for GIS. But even thematic maps come in many different scales. A good rule of thumb is that the larger the map scale, the smaller the area covered and the greater the detail. Larger scale maps are generally better for your GIS activities because they provide the largest amount of detail.

If all maps were of the same highly detailed scale, you wouldn't have to think about what scale a particular map uses. Unfortunately, you find many maps at many different scales. In fact, most GIS databases contain a lot data from differently scaled maps. Each map has more or less detail, more or less positional accuracy, and more or less information content. When you're looking on a map for data to include in a GIS, the cartographer has already decided what to include and what to leave out, how to symbolize features, and how much to generalize the shapes used on the map.

Using a wide range of data

Thematic maps come in all sorts of sizes and shapes, and they can cover all kinds of different topics. Think about all the things that actually occur on the Earth, and you start getting some idea about the number of potential thematic maps. Here's a short list of some of the kinds of maps that contain socioeconomic information and how you might begin to learn to recognize and apply these patterns for business:

✔ **Median family income:** If you're trying to start a business that sells expensive jewelry, you probably want to place your store in an area that has a high median family income.

✔ **Education level attained:** If you sell technical books, you want to know where people with college degrees live so that you can locate near them.

✔ **Mean number of children under 5 years old:** If you want to sell baby clothes, you ideally want to establish your store close to large concentrations of people that have young children.

This set of examples uses a useful concept in geography — all things are related to each other in geographic space, but close things are more related than far things.

Comparing thematic maps

All things are related to each other in geographic space, but close things are more related than far things. This concept is used frequently in GIS when thematic maps are compared to see what things occur near each other. For example, you can use these maps to do a variety of analyses:

✔ Earthquake hazards

✔ 100-year flood zone

✔ Soils with high shrink-swell clay

If, for example, you can locate the zones with the highest frequency of earthquake, your insurance company can decide whether to issue earthquake hazard insurance, or your builder might have to use some earthquake-proof building techniques to allow for that type of activity. Insurance companies that issue flood insurance need to know whether their potential customers live within an area that's likely to flood frequently. The building contractor would find a soils map that shows high shrink-swell clay useful so that he or she knows whether to add pylons to reinforce that new skyscraper to keep it from becoming the next Leaning Tower of Pisa.

The more detail, the more complete the information, and the more precisely located the geographic features, the better the analytical results you can get. You might think that the poorest quality map determines the quality of your analysis. This *weakest link hypothesis* makes sense on the surface, doesn't it? In cases where each map layer is of equal importance, this hypothesis might be true. However, in most GIS analysis that you do, the different map layers aren't equally important. Maps that are poorer in quality may still add value to your GIS as long as their lower accuracy does not affect the overall quality of the analysis.

Working with Projections and Datums

Except for members of the Flat Earth Society, everyone accepts that the Earth is roughly spherical. This spherical shape has some major drawbacks for the mapmaker who's faced with producing a flat map that correctly represents the shapes, angles, distances, and sizes of objects on the Earth. In simple terms, the conversion from a sphere to a flat map is always going to result in some distortion. Peel an orange and try to flatten the peels on a table; now you have a visual idea of the problem that the mapmaker faces.

Picking the right projections

Figure 2-4 shows some of the many ways a map can look. Map *projections* — the process of converting the spherical Earth to a flat surface — come in many different types, from contiguous to interrupted, from those that look like photographs of the Earth to those placed on cones or cylinders. In fact, cartographers often refer to the three families of map projections as planar, conic, and cylindrical, which describe the surfaces onto which the surface of the Earth would be projected (hence the term projection) if a light bulb could be put in the center of the Earth.

Figure 2-4:
A map can represent a place in many different ways, each of which produces some form of distortion.

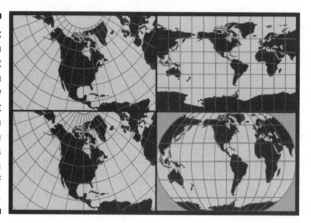

The curved surfaces of the Earth just don't cooperate when you try to make them flat. The mapmaker has to figure out which of the map's properties he or she most wants to preserve. For maps designed to measure distance, the mapmaker might choose to preserve distance on the map. Another map might be used to determine directions, which requires that shapes be preserved. Anyone trying to measure areas or changes in areas wants a map that preserves area.

The good news is that you don't have to know all the gory details about how the mapmaker preserves certain properties to work with GIS — but you do want to know which properties are mostly preserved.

When working with GIS, pick the map projection that best represents the properties you want preserved when you create output maps from your analysis. For example, if your analysis involves changes in area, a map projection that preserves area is the best type to use to display your results.

Most high-end GIS software has the capability to convert back and forth from one projection to another. Most have more map projections than you'll ever use. But every time you change from one projection to another, the software rounds off numbers and makes tiny errors in position. If you convert projections repeatedly, these tiny errors in position start to add up. Eventually, you can add enough error to have a negative impact on your analytical calculations.

Good projections depend on accurate datums

The impact of converting projections, like with many other GIS-related transformations, depends on scale. Small scale maps will be affected more than large scale maps. Having an accurate representation of distance and area measurements in any projected map depends on having accurate measurements of the spherical Earth. The science of *geodesy* deals specifically with these measurements.

Like with any measurement, you need to have a place to start, a baseline. In geodesy, such baselines for measuring the Earth are called datums. In simple terms, the *datum* is a model of the Earth's shape that allows geodetic scientists and surveyors to accurately measure the Earth and place the origins and orientation of the coordinate systems.

Each of the hundreds of datums has different measurement units and methods for measuring the Earth. Some datums start by measuring the Earth's diameter in different directions, others focus on flatness ratios (ratio between the polar and equatorial diameters), and still others are based on measuring the Earth's circumference at different places Because each datum uses different methods and different models of the Earth (which are called *reference ellipsoids*), the maps associated with a datum depend on these properties. Table 2-2 gives you an idea of the possible datums you might encounter in your work. You can find a quite exhaustive (or is that exhausting?) list of datums online at

`www.colorado.edu/geography/gcraft/notes/datum/datum_f.html`

Table 2-2	A Few Datums Used in Map Projections	
Ellipsoid	*Semi-major axis*	*1/flattening*
Airy 1830,	6377563.396	299.3249646
Modified Airy	6377340.189	299.3249646
Australian National	6378160	298.25
Bessel 1841 (Namibia)	6377483.865	299.1528128
Bessel 1841	6377397.155	299.1528128
Clarke 1866	6378206.4	294.9786982
Clarke 1880	6378249.145	293.465

(continued)

Table 2-2 *(continued)*

Ellipsoid	Semi-major axis	1/flattening
Everest (India 1830)"	6377276.345	300.8017
Everest (Sabah Sarawak)	6377298.556	300.8017
Everest (India 1956)	6377301.243	300.8017
Everest (Malaysia 1969)	6377295.664	300.8017
Everest (Malay. & Sing)	6377304.063	300.8017
Everest (Pakistan)	6377309.613	300.8017
Modified Fischer 1960	6378155	298.3
Helmert 1906	6378200	298.3
Hough 1960	6378270	297
Indonesian 1974	6378160	298.247
International 1924	6378388	297
Krassovsky 1940	6378245	298.3
GRS 80	6378137	298.257222101
South American 1969	6378160	298.25
WGS 72	6378135	298.26
WGS 84	6378137	298.257223563

Your GIS software needs to know what datum you're using for each set of map data that you put into your database. Attaching your coordinates to the wrong datum can result in location and measurement errors. You'll commonly begin with maps that use different datums. To make your GIS data from maps of the same area correspond to each other, they all need to be converted to a common datum.

When loading up your GIS, be sure to use the correct datum for each map source as you add it. Also, convert all your map data to a common datum when you work with more than one source map at a time.

Working with Coordinate Systems and Land Subdivisions

No matter what the theme, projection, or scale, you need to find your way around on your maps — and that's what coordinate systems help you do. On a globe, you use the latitude-longitude system. On flat maps, you have to convert this geographic grid to something a bit more practical. Ultimately, all coordinate systems are based to some extent on the geographic grid.

But after you take a spherical object such as the Earth and flatten it, you need to have a grid that conforms to this new surface. In other words, after you project the spherical Earth onto a flat surface, you need to put a set of direction lines on that flat surface which conform to the distortions the projection just produced. So, coordinate systems are often linked to specific types of map projections.

Meeting the Universal Transverse Mercator (1 know you want to)

You don't need to know the details of all the different types of coordinate systems to be good at GIS. You do need to know the basics of how coordinate systems work, though. One of the many possible coordinate systems that you may encounter in GIS is called the UTM system. UTM stands for *Universal Transverse Mercator,* which is the most commonly used system.

This system divides the Earth from latitude 84° north and 80° south into 60 numbered vertical zones, each 6 degrees of longitude wide.

Each UTM zone has two places from which measurements begin (origins):

- The equator
- 100,000,000 meters south of the equator (80° south latitude)

Dividing the Earth by using UTM has two big benefits:

- You can measure stuff in positive numbers east and north from one of the origins, regardless of which hemisphere you're in.
- You never have to use negative numbers.

Figure 2-5 shows a diagram of the UTM coordinate system. The zones are numbered, starting at the 180th meridian, in an eastward direction. Each zone is divided into rows (or sections) of 8 degrees latitude each, with the exception of the northernmost section, which is 12 degrees, allowing the system to cover all the land of the northern hemisphere. You can isolate small sections of the Earth because you locate them by a number/letter combination. With the exception of the northernmost group, each of these sections occupies 100,000 meters on a side and can be designated by *eastings* and *northings* (that is, positive measurements eastward and northward of the origin) of up to five-digit accuracy (1-meter level resolution).

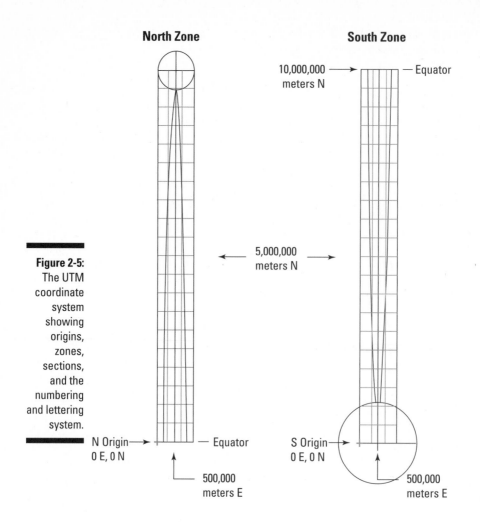

Figure 2-5:
The UTM coordinate system showing origins, zones, sections, and the numbering and lettering system.

Measuring the land

Coordinate systems help you measure geographic features, as well as find your way around and locate things on a map. One major feature you'll likely encounter in your GIS work is land ownership, and you'll find many systems that measure and record this feature. I describe a common one used in the United States to help you understand the relationship between land measurement and the coordinate system.

The U.S. Public Land Survey System (PLSS) was established in the late 1700s to divide up the ever increasing lands that the U.S. Government acquired. The PLSS is similar to a rectangular coordinate system, but it's not directly linked to any particular map projection.

Maps as questions, as well as answers

Most map readers assume the map is accurate, which isn't always true, depending on the scale. More importantly, though, most believe that the map answers the question, "Where is _____?" You fill in the blank. Well, I'm not going to argue with you there — the map really does display (within its graphic and scale limits) locations and distributions of all sorts of geographic features. But that's just the beginning. The geographer and the GIS professional often see a map as just the beginning, rather than an end.

Let me help you understand where these folks are coming from. The GIS is designed to input data from maps, and then compare the locations and distributions of geographic features from one map to another. Well, would you add map data into the GIS if you didn't think that you could use the data for analysis? And you probably select your maps with particular questions in mind. Because you're using maps, it's also a safe bet that your questions revolve around the locations and patterns of the geographic features you choose to include.

So, for example, if you're a crime analyst, you might have a map of auto thefts for your city. This map answers the question, "Where are the auto thefts?" But, if you needed to know only where thefts occur, you wouldn't need to put it into the computer. You have questions about the locations and distributions. You might want to know some of the following:

✔ Do the auto thefts occur more in some areas than in others — does your city have auto theft "hot spots"?

✔ Do the number of cars stolen increase at any particular time of day, day of the week, month, or year (is there a temporal pattern)?

✔ Do other distributions seem to match that of the auto thefts? Or, stated differently, do spatial correlations exist between auto thefts and other factors, such as neighborhood type, access to major highways, or gang activity?

The map is both an *answer* and a *question*. When you look at a map, especially if you're considering including it in a GIS database, you should start asking questions about the distributions and patterns that you observe. The question is always the same — "What explains this distribution?" GIS gives you tools to answer that question. And if you can explain a distribution, you can begin to predict future distributions and make plans accordingly.

So, if you're a transportation planner and you notice that the traffic counts are increasing in a particular neighborhood, your analysis might show that this traffic increase is very strongly related to population increases in certain neighborhoods. Your next step might be to build a predictive map of future transportation while the population grows. You can then use that prediction to suggest improvements to the transportation system. In a way, GIS makes you a really great prognosticator.

This land ownership recording system is based on the acre, which is 1/640 of a square mile (a *section*). Thirty-six sections are grouped together into townships and numbered in serpentine fashion from upper-right (section 1) to lower-right (section 36). A land records specialist can subdivide these chunks of land into halves and quarters so that he or she can describe the land holdings by these fractions. For example, the PLSS system describes the plot of land you own as something like NE1/4 SE1/4 N1/2 Sect. 23, T144, R35E.

All these land portions are part of a larger coordinate system of horizontal and vertical lines. Distance is measured north and south (township) of starting lines called *baselines,* and east and west (range) from starting lines called *principal meridians.* These township and range lines allow for the legal descriptions of land.

The notation that identifies a parcel of land starts with the smallest subdivision of land, then gives the section number, and then township and range numbers. A legal land description based on this system includes the state name, principal meridian name, township and range designations, and section number. Figure 2-6 shows a breakdown of the PLSS system.

Such land subdivision systems enable you to build GIS databases that have land ownership correlated to other factors, such as zoning restrictions or price per acre.

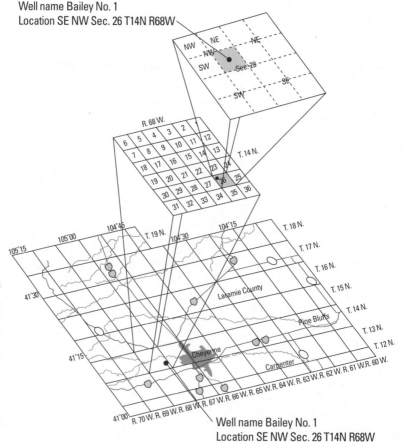

Figure 2-6:
Subdividing
land by
using the
PLSS, or
township
and range,
system.

Chapter 3

Reading, Analyzing, and Interpreting Maps

*P*eople who explore GIS are often most astonished by the wide variety of content-specific (thematic) maps available. You need to have a working knowledge of the many varieties of maps and the types of symbols they use so that you can determine the best way to input data, as well as know the types of maps you use for a given application.

While you become familiar with maps, especially thematic maps, you begin to identify patterns in the symbols on the paper. Spotting patterns isn't as easy as you might think because those abstract symbols represent sensory data contained on the map. The experience is akin to finding your way around a new town — you get the hang of it while you press further into your exploration.

This chapter shows you how to identify patterns within the sometimes complex map symbology. Also, you can find out how to ask questions about the patterns you identify, answer those questions through GIS analysis, and ultimately make decisions based on your analysis.

Making Sense of Symbols

Many types of maps are available, and each type has a unique set of symbols. *Symbols* are lines, objects, or pictures on the map that represent real objects on the ground. With so many map types, each with its own set of symbols, you can easily become confused when trying to interpret the maps. You can make the process easier on yourself by looking at the dimensions and the types of measurement scales available for mapping and GIS operations, as described in the following sections.

Categorizing the space on a map

Categorizing the space represented on maps helps make symbology easier to understand. The types of space that occur in maps are

- ✔ **Points:** Considered zero-dimensional because they take up no space at all. Points show up as symbols representing the locations of objects such as towns, wells, or churches. These symbols might be dots for towns, a tiny caricature of a well for wells, or a cross for Christian churches.

- ✔ **Lines:** One-dimensional (measured by length). Lines might be streets, roads, or railroad tracks symbolized by different types or colors of lines on the map.

- ✔ **Areas:** Two-dimensional (measured by length and width). Areas might be color patterns on city street maps that indicate different neighborhoods or suburbs, or they might be patches of blue to show the location of lakes.

- ✔ **Volumes:** Three-dimensional (measured by length, width, and depth or height).Volume spaces are important when you want to identify something such as a buried coal seam. Volumes show up as area symbols on a map.

When it comes to map symbols, volumes and surfaces are treated pretty much the same. Unlike volumes, however, surfaces are thought to "blanket" something, rather than include what's inside. For example, if you're interested in the vegetation covering a mountain, you're really interested in only the elevation's surface. But, if you're interested in how much coal is located underground, you need to know not just the surface of the ore seam, but also its whole volume because you'll be extracting that volume.

Four types of objects are symbolized on your map — points (zero-dimensional), lines (one-dimensional), areas (two-dimensional), and volumes (three-dimensional).

Understanding levels of measurement

All map data, including point, line, area, and volume data, come in one of four levels of measurement:

- ✔ **Nominal:** Objects are often identified by their name — a church, a house, an interstate highway, rail lines, the range of a plant or animal species, or a parcel of land. This level of measurement is called *nominal,* meaning *named.* You can't compare nominal measurements to each other. The comparison can be literally apples and oranges.

- ✔ **Ordinal:** You might also categorize things by a general category of size, such as a small, medium, or large house, road, or land parcel. Such objects are measured at the ordinal scale. The *ordinal scale* characterizes things by their rank or order in a sequence, so the name ordinal refers to the ordering or ranking.

- ✔ **Interval:** If you have detail and can actually measure things by intervals, you might use the interval scale. The *interval scale* also places things in order, but you can measure the relative difference by exact intervals. For example, the number 10 is 90 units lower than 100.

Because points are technically considered to have no dimensions, measuring them at the interval scale may seem, well . . . pointless, but that's not necessarily the case. The point may represent something that's measurable at that scale, such as an average soil temperature in degrees Fahrenheit.

You can't make ratios out of interval scale objects. So, if a soil is 32 degrees Fahrenheit in one place (A) and 16 degrees Fahrenheit in another (B), you can say that A is 16 degrees warmer than B, but you can't say that A is twice as warm as B. The starting point for Fahrenheit was chosen for convenience rather than to represent an absolute starting point where the temperature is at the lowest possible level. Dates on calendars are also interval because different calendars have different starting points (for example, the Mayan calendar versus the Gregorian calendar), and the starting points aren't based on when time actually began.

- ✔ **Ratio:** Items that have an absolute starting point, such as population (where zero is the lowest number you can have), are measured at the ratio scale. The term *ratio* is used because you actually can make comparison ratios out of such things. For example, a city (A) that has a population of 2,000,000 people has twice the population of another city (B) that has 1,000,000 people — the ratio here is 2 to 1 or 2:1. Other examples of ratio data include the size of a parcel of land, the length of a border, or the volume of a buried ore body.

GIS incorporates four primary scales of geographic data measurement —
nominal (named, non-comparative), ordinal (ranked, relative sizes), interval
(measured by increments, but with an arbitrary starting point), and ratio
(measured in increments with an absolute starting point, allowing for ratio
comparisons).

Understanding the relationship between symbology and data measurement

Figure 3-1 shows a table that illustrates the relationships between each scale
of data measurement and the symbols depicting geographic features. Each
table entry includes an example or two of the type of geographic features
that you might run across in your GIS activity.

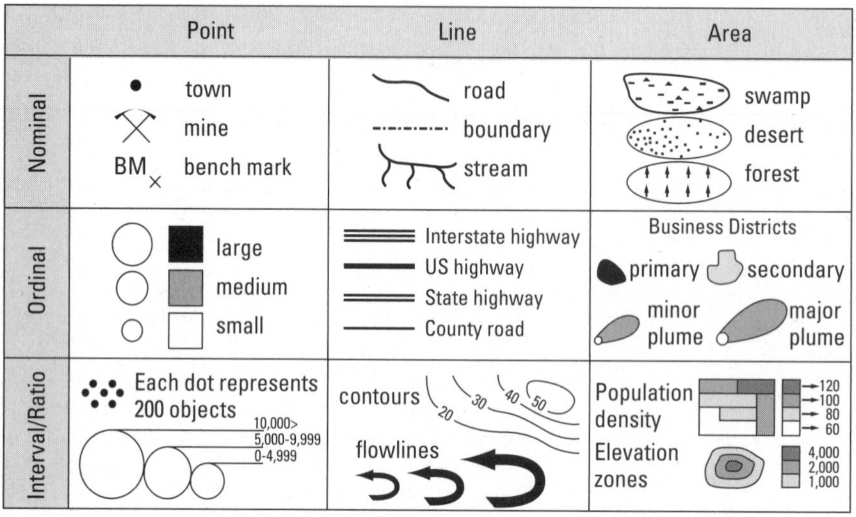

	Point	Line	Area
Nominal	• town ✕ mine BM✕ bench mark	road boundary stream	swamp desert forest
Ordinal	⬤ large ◯ medium ◯ small	Interstate highway US highway State highway County road	Business Districts primary secondary minor plume major plume
Interval/Ratio	Each dot represents 200 objects 10,000> 5,000-9,999 0-4,999	contours 20 30 40 50 flowlines	Population density →120 →100 →80 →60 Elevation zones 4,000 2,000 1,000

Figure 3-1:
Geographic
features and
how they're
measured.

In Figure 3-1, you can find some symbols used for various features on a map:
point (zero-dimensional), line (one-dimensional), and area (two-dimensional).
Along the left side, the list of measurement scales combines interval and
ratio. Although these two data measurement scales do have their differences,
the combination makes sense because they share identical symbology.

Working with nominal features

The data measurement level shown at the bottom of Figure 3-1 is nominal.
The figure shows typical geographic objects that you encounter on a map,
including a town, a mine, a stream, and a forest. A picture or graphic symbol

represents each of these features on the map. These features and many more like them are *non-comparative,* which means you can't compare them to each other directly. You can't easily compare Oakmont, the town, to the Black Hills mine.

Other non-comparative features include

- ✔ **Line features:** Streets, rail lines, and boundaries, for example, are unique entities and can't be compared to one another. Geographers say that they're not the same *kind* of feature.

- ✔ **Areas:** A swamp, the range of a wild species, the land owned by the federal or state government, or the type of zoning for a particular parcel of land are also unique and can't be compared.

- ✔ **Volume features:** Water aquifers, hills, and buried ore bodies all take up volume and are named on maps. You can acknowledge that they exist and give them specific names, but you can't compare ore bodies to aquifers.

Depicting features that use the ordinal scale

You can add *ordinal* (ranked) attributes to points, lines, areas, and volumes. Inhabited regions have names, and you can also rank them as hamlets, villages, towns, and cities. This ranking indicates that the features aren't only the same kind of feature — inhabited places — but they're also different sizes of inhabited places.

Geographers call these distinctions *comparisons of kind,* and apply such ranking as follows:

- ✔ **For line data, such as roads:** You may want to depict dirt roads, single lane highways, superhighways, and so on. These lines are comparable because they have attributes that are relative in size (ordinal).

- ✔ **For volume data, such as an ore body:** You can categorize the ore quality as low-, medium-, or high-grade ore.

Comparing data using the interval and ratio scales

Although the symbols used to represent interval and ratio scales are identical, mapmakers have many options for designing these symbols. To indicate different sizes on the scales, the cartographer can change the size of symbols. How much the symbols vary in size is directly related to the size of the features that the symbols represent, which is why cartographers call such symbols *graduated symbols.* So, each symbol can be *graduated* (or sized) to the feature it represents. Easy squeezy.

Not so fast — there's a catch. Consider this example: The United States has about 3,000 counties. If each county has a map area that contains, say, only 30 towns and cities (in reality, most counties probably contain many more towns and cities than 30), you have 90,000 towns and cities that all need symbols. Even if you map only one town for every third county, you still have 1,000 symbols. If each town has a different population, a dedicated cartographer calculates the relative size for each one of those 1,000 symbols. Just imagine the frustration for a map reader trying to identify the differences among so many symbols.

The good news is that you can find a better way to select symbol sizes so the reader will be able to recognize them. You can group these point symbols to show that features have various sizes by assigning them to *classes* which break up the range of values into a small set. For example, the population of cities can be grouped into classes with limits of 0 to 10,000, 10,001 to 100,000, 100,001 to 1 million, and over 1 million. Cartographers have a lot of methods that they can use to determine the class limits for point symbols. More importantly, they can apply these same methods to line, area, and volume data that are measured with the interval/ratio scales. Here are some examples:

- ✔ **For line features:** Imagine using different-sized lines to represent the levels of flow in a river or stream (see Figure 3-1).

- ✔ **For area features:** Consider showing the potential area that a hazardous spill might occupy as a set of differently sized tear-drop shaped symbols.

- ✔ **For volume features:** A cartographer can represent the sizes of ore bodies as a set of irregularly shaped area symbols that increase in size, or even symbols shaded to look 3-D.

Cartographers spend a great deal of time creating symbols for point, line, area, and volume features. They've developed a host of methods for standardizing the symbols and determining class limits. The good news is that this work has been done for you. The so-so news is that the map data grouped together in this way are less accurate than the raw data used to compile the maps.

Recognizing Patterns

One cool thing about maps is that the symbols represent a scaled-down version of real geography. As much as possible, the map's symbolic objects, features, and background are distributed and located in ways that closely resemble the locations and distributions of real objects. This aspect of a map is important because patterns of geographic features are — more often than not — related to an underlying process and, as a result, represent cause and effect.

The processes that create location and distribution patterns often took place in the past and are typically still in play during the present. By extension, the processes operating today will continue to operate in the future. Although these statements are generalizations — subject to changes in speed, and the continuation or secession of the underlying processes — making this assumption allows you to undertake the description, explanation, and prediction of patterns. In this way, you interpret the patterns — but first, you need to recognize that patterns exist.

When you identify patterns, you look for a degree of predictability in the arrangement of the objects you're interested in. For example, if you see a map showing a group of trees that are dying from some disease, you can predict that the adjacent trees are likely to become infected and die eventually.

Each person interprets patterns differently, based on experience, background, and a variety of other factors. To make the best use of GIS information, the trick is to notice patterns that you may not be used to seeing, such as patterns of trees, houses, roads, rivers, or any other features you encounter.

Identifying random distributional patterns

People see spatial patterns based on the separation between objects that they're observing, meaning they notice how close together or far apart the objects are. Some objects, such as dandelions in your yard, occur pretty much at random (like in Figure 3-2). This type of distributional pattern occurs because of the process that spreads the dandelion seeds. The seeds are spread by the wind, carried on animals — or even people — and spread in other ways that produce a random distribution. The spatial pattern made by these dandelions reflects the widely varying distances between individual plants. Some dandelions are close to each other (within inches), some are far away (several yards), and others are somewhere in between (a few feet).

Figure 3-2:
Dandelions create a random distributional pattern.

Finding clustered distributional patterns

Certain features occur in clumps, or *clusters*. In such cases, similar features tend to be very close together (in a cluster), and another cluster of these same features may be far apart from the first cluster. In a clustered distributional pattern, a significant number of features tend to cluster together in portions of your study area. For example, crimes tend to occur in clusters, called *hot spots,* around specific neighborhoods (as shown in Figure 3-3). These hot spots reflect the processes that facilitate crime, such as the availability of victims or the location of criminals. You might, for example, expect to find clusters of jewelry thefts in locations where wealthy people who have expensive jewelry live. After you start seeing spatial patterns, you see them everywhere — and you then begin to naturally make the link between patterns and the possible causes and ramifications of the patterns.

Be careful when making interpretations of patterns. Maps that show clusters of tornados might tell you more about the number of people who are around to observe those particular tornados than it does about your weather patterns. You can also be misled into making the incorrect conclusion that the more people in an area, the more tornados occur. The population is related to tornado sightings but hopefully not to the cause of tornados.

Figure 3-3:
A clustered distributional pattern appears in a map of crime hot spots.

Dona Ana

Observing uniform distributional patterns

The type of pattern that's the easiest to observe is called *uniform distribution*. An orchard (like the one in Figure 3-4) is a classic example of a uniform distribution. Each tree is spaced almost exactly the same distance from each adjacent tree. This type of pattern rarely occurs naturally. It's pretty obvious that the pattern is driven by some form of human action — in this case, planting.

Figure 3-4:
Humans create a uniform distributional pattern in an orchard.

Seeing patterns among dissimilar features

You may see geographic features present in some places (whatever their spacing) and absent in others. The sizes, shapes, orientations, and juxtapositions of these features, compared to other map features, provide strong evidence that a spatial relationship exists between unlike features. Consider these examples:

- ✔ A map may show places that have wetlands and places that don't. And you know that wetlands often occur in places that feature a depression in the landscape.

- ✔ You may notice that commercial development seems to occur in only selected parts of the city. In this instance, you know that many city codes and zoning regulations control the locations of commercial enterprises.

- ✔ You observe that a species of plant or animal shows up in one place and not another. Typically, plants and animals adapt to differences in the physical environment at various locations.

Each of these distributions indicates something about the underlying processes (cause and effect), in the same way that random, clustered, and uniform spacing do (as I discuss in the preceding sections). After you recognize the patterns, you can start thinking about the relationships between patterns and associated processes and how you can exploit that knowledge.

Describing patterns with linear features

Linear features provide both locations and distributions, but because they're often connected in a variety of ways, you can describe some additional patterns — including connectivity, circuitry, and linkages — that you don't find in other types of features.

Many linear features connect to each other in a series of pathways, like in the example of road networks and railroad lines. If you look at a map of the major highways of the United States, for example, you notice a pattern immediately. Some places, especially highly populated locations (such as the eastern seaboard), have beaucoup highways with tons of connections and many ways of getting from one place to another. Other places, such as the Great Plains, have way fewer highways with much less connectivity. These different road systems exhibit patterns of road networks based on the number of connections, not just on the number and proximity of roads.

Some road networks and highways allow you to go around obstructions, rather than just through them. These networks form closed loops, or *circuits,* just like in an electrical circuit. When road networks have circuits, traffic has an alternative because it can flow around obstacles that drivers might want to avoid. Rail lines use circuits so that trains can more easily pass other approaching trains. Perhaps one of the most common forms of transportation circuit is the U.S. Interstate Highway bypass (or outerbelt) that surrounds major metropolitan regions and allows traffic to avoid the congested areas of the internal part of the city (see Figure 3-5). If you see a concentration of circuits, such as outerbelts, you can be fairly certain that the population is high in that location because circuits are designed to allow a lot of vehicles to travel around the densest part of the city where traffic is congested and slow.

Understanding the repeated sequence of shapes

You can recognize patterns when items are arranged to form a repeated sequence of shapes. A simple example of such a pattern is clothing that features a pattern, such as a herringbone suit or a plaid shirt, rather than a solid color. Such patterns occur in linear objects on the Earth, as well.

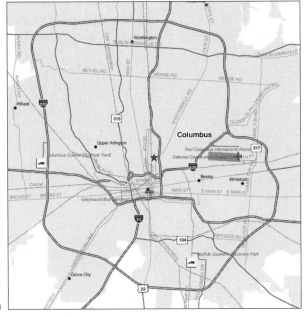

Figure 3-5:
A highway
outerbelt
allows
traffic to
flow around
the city.

One well-studied set of patterns has to do with how rivers and streams connect to form different designs. Although you can find different terms to describe the following types of patterns, the ideas remain the same. Figure 3-6 shows a thumbnail of each of the following patterns:

- **Dendritic:** The easiest one for me to remember is called *dendritic* and looks like the branches of a tree. These branches (called *tributaries*) go out in all directions and seem to have a mind of their own. The flat terrain and absence of strong rock formations allow the water to go in somewhat random directions.

- **Radial:** Another common pattern that streams take is called radial. A *radial* pattern looks like a dendritic pattern, except that all the streams flow outward, away from a center, like the spokes of a wheel. A radial pattern occurs because a hill (with its higher elevation) at the center of the pattern works with gravity to cause the streams to flow outward and downward from the top.

- **Centripetal:** The opposite of radial patterns, *centripetal* patterns occur when a low spot or depression affects the flow. The radial pattern is reversed, and the streams flow toward the center.

- **Parallel:** *Parallel* and *sub-parallel* streams run along the gentle slopes that result from either natural topography or from land manipulation because of road or mining construction activities.

✔ **Trellis:** *Trellis* stream patterns result from rock strata that's jointed, exposed, and folded from geological forces acting over time. Trellis patterns resemble the street patterns in neighborhoods loosely organized along a grid.

✔ **Rectangular:** *Rectangular* stream patterns show strong right-angle turns and are often the result of cross-cutting joints in the underlying rock. This pattern also occurs in some man-made neighborhoods where drainage ditches are strongly oriented along a grid.

✔ **Annular:** Annular stream patterns occur in areas which have a dome that is eroded. The erosion forms a series of circular rings that have fractures at right angles to the rings, and the streams are forced to flow in the interrupted portions of these circles. The term *annular* comes from the geometric form *annulus,* which is a ring.

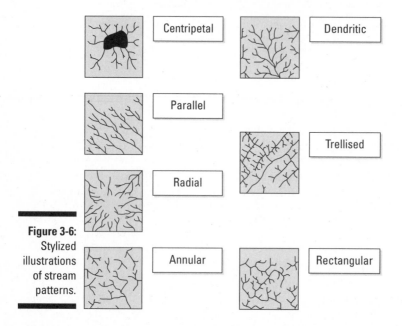

Figure 3-6:
Stylized
illustrations
of stream
patterns.

Centripetal

Dendritic

Parallel

Trellised

Radial

Annular

Rectangular

Analyzing and Quantifying Patterns

Recognizing that patterns exist is a great start for any GIS user, especially the analyst. Every pattern occurs for a reason, even if the reason is a random one. You analyze patterns to determine whether the process that created them was random, clustered, or uniform (see the section, "Recognizing Patterns," earlier in the chapter). When you know the processes underlying the pattern, you can begin to take that explanation and make predictions about future patterns based on your knowledge of the presence, absence, or intensity of that process. For example, if you notice that a portion of a crop isn't doing well, you might find that the soil in that area is low in nutrients — perhaps because of overplanting or under-fertilization.

Knowing your geometry and patterns

To perform a useful analysis of the patterns in your map, you must recognize the geometry and patterns that you encounter. You need to be able to do three basic things with the patterns:

✔ Accurately describe the patterns in a way that the layperson can understand.

✔ Quantitatively compare and contrast your descriptions of the patterns to those of other features and feature sets.

✔ Analyze the quantities that you identify to determine a measure of their pattern or shape, and to provide a baseline for changes that might take place over time.

Analyzing patterns in the real world

The recognition, quantification, and analysis of spatial patterns together carry one important characteristic. They give you the power to interpret what those patterns mean and what importance the patterns might have (if they have any). This chapter includes a few examples about the significance of some patterns, but the list of pattern types and causes is nearly limitless:

✔ Wildlife specialists want to know the locations, extent, and movement of animals so that they can examine the natural or human causes of changes in animal population or habitat.

✔ Urban planners examine street patterns, utility patterns, urban density patterns, regional population changes, and patterns of traffic to control growth so that everyone can move about; have heat, light, and water; and live apart from noisy or polluting industries.

✔ Military analysts need to understand the locations and distributions of troops and armament so that they can deploy their own forces effectively.

✔ Criminologists study distributions and concentrations of crime so that they can protect the public.

✔ Utility companies need to examine their distribution networks for wear and damage from weather and age.

✔ Businesses analyze the distribution and composition of their potential customers so that they can decide where to open new branches.

✔ Emergency care personnel need to know the connectivity of street patterns and the patterns of traffic at different times of day.

GIS is designed to do much of the analysis automatically. The software antici-pates that you'll need these types of values for future analytical techniques. All pattern descriptions require some basic GIS operations, however. To analyze patterns of numerous objects, the software needs to be able to count objects, determine the absolute locations of the graphical entities (the actual points, lines, and polygons), and measure objects and the spaces between them. These techniques are the same ones that you use to calculate geom-etry and orientation (see Chapter 11 for more on these calculations).

Using GIS software for the analysis

Your GIS software knows which values you need for your analysis because when you put your spatial data into the computer, you tell the software exactly where each point is and how that point relates to the feature's real position on the surface of the Earth. When you turn the software on and load up the database, most programs immediately retrieve all the points, lines, polygons, and surfaces that you put in the database and report on the following calculations:

✔ The number of points, lines, and polygons (areas) for the theme you load.

✔ The individual polygon areas and the total area for the theme that you loaded.

✔ The total lengths of linear objects of a particular category.

In short, the software automatically calculates much of what you need to describe patterns and spatial geometry.

So, what's left for you to do? Well, the GIS can't anticipate everything you might want to calculate or analyze, so you do have to make some deci-sions. You can find so many new tools for analyzing that you may feel over-whelmed. I try to help you through the decision-making process by reviewing a few common descriptive analyses in the following sections. You can find more advanced methods, but these simple ones are quite powerful.

Determining the type of pattern

Figure 3-7 shows the distribution of tornados in Illinois from 1950 to 1998. This distribution is clearly not a regular distribution. But deciding whether the distribution is clustered or random is a bit more difficult. Even when a distribution pattern visually appears to be one or the other, you need some way to identify the pattern for sure.

In the clustered condition, some points are closer together than others. Put another way, the points in the clusters have many close neighbors, whereas the remaining neighbor distances are much farther apart. The mathematical technique that the GIS uses to make this determination is called, not surprisingly, the *nearest neighbor statistic*.

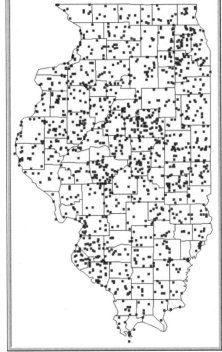

Figure 3-7:
GIS can determine whether this tornado pattern is clustered or random.

To determine whether a pattern of points is regular, random, or clustered, your GIS software follows these steps:

1. **Find the random nearest neighbor distance by dividing one by twice the square root of the density of points.**

   In the tornado example, you realize when performing your analysis that the number of tornados for Illinois isn't the same from one season to the next or from one year to the next. You need a measurement of the average nearest distance that would occur if the distribution were random. That measurement is based on the *density* (number of points per unit area) of the distribution, which you identify with this simple equation:

   Random nearest neighbor distance = 1/(2 × square root of the density)

In this case, density is the number of tornados per square mile. The average nearest distance changes with the density but always gives you the nearest neighbor distance for a random pattern.

2. **Calculate the average pair-wise nearest neighbor distance of your points.**

   The *pair-wise* part of this term indicates that each point (denoting a tornado location) is matched with only one other point as its nearest neighbor. And this statistic is pretty easy to calculate. The GIS software finds the *x*- and *y*-coordinates for each tornado location, measures the distance to every other location, and determines which location is the nearest neighbor. After collecting this data, the program calculates the average of these *pair-wise nearest neighbors*.

3. **Divide the calculated average pair-wise nearest neighbor distance by the random nearest neighbor distance to compare distances for randomness.**

   After the software calculates the average pair-wise nearest distance, you need a way to compare the standard for random distributions with other types of distributions — regular and clustered. You just have to compare the measured nearest neighbor distance to the random distribution by dividing the measured (calculated in Step 2) by the random (calculated in Step 1).

4. **Compare the number to 0 and 1, where 1 is perfectly random and 0 is perfectly clustered.**

   Numbers larger than 1 indicate a more regular pattern.

If you calculate the mean of all the nearest neighbor distances and divide that number by the random nearest neighbor distance — and you get a value of 1.0 — the measured nearest neighbor distance shows a perfectly random distribution. If the value you get is 0.0, this measurement is perfectly clustered (meaning as clumped together as you can get). If you get a value greater than 1, your distribution becomes more regular (more dispersed) as the number grows larger. No absolute value exists for determining perfectly regular distributions because the values you get depend on the density of points.

You can determine whether the values that you get from your analysis occur by chance or whether they're truly meaningful, but you may want to leave that job for the statisticians in the crowd. You need to know only the nature of the distribution.

The nearest neighbor analysis is an effective tool for determining the pattern of point objects, but you can also use this analysis to evaluate the distribution of line objects, and even area and volume objects. You just need to let the software calculate the center of each line — or each polygon or volume feature — and then do the same calculation that you perform for point objects (for example, nearest neighbor). You use nearest neighbor statistics to make sure that what your eye has already told you is real and not a figment of your imagination.

Identifying even more patterns

Some calculations can help you identify different types of patterns. You can use certain techniques to determine the average direction of linear objects, such as tornado paths, fallen trees, boulder distributions, shelterbelts (protective barriers such as windbreaks), and many other objects. Without going into the trigonometry, it simply gives you a summary statement of the direction in which these events and objects occur. This calculation might, for example, tell you the direction of wind during storms, the movement of a glacier while it leaves debris behind, or the way that farmers line up planted trees to protect their crops from the wind.

Beyond these calculations of position and dispersion, you can find techniques for calculating the connectivity or circuitry of networks, the ratios of streams branching, the sinuosity of streams, the isolation of forest patches, and many more available inside the GIS software. In most cases, you can find these techniques simply by knowing the appropriate terms describing the condition of the geographic features you wish to examine (connectivity, isolation, and so on) and what they mean, and then searching for them in the help menu. The best part is that most GIS software not only tells you how the commands work, but it also frequently provides short tutorials that explain what the statistics do.

Interpreting the Results and Making Decisions

Maps show many types of patterns and distributions. As you read more maps, you become more familiar with them and often recognize that many of the patterns are distinctive in their configuration, distribution, or location. The GIS provides you with a mathematical means to quantify and categorize these patterns so that you know they really exist. Although wielding such quantitative power is cool, you don't use GIS just for fun unless, like me, you *really* like patterns.

Each pattern exists because of underlying forces, whether those forces are natural or human, random or focused, static or dynamic. After you become familiar with the idea of seeing and recognizing patterns for your own specialty, you next need to use your specialized knowledge to identify the reasons they exist. A process creates every pattern, and every pattern has an effect on the process that created it. As trees reseed themselves in a single plot of land, the density of trees increases. As the density of trees increases, the competition for space increases, and some trees die, thus reducing the density.

When you analyze your GIS data, follow these steps:

1. **Recognize and acknowledge the existence of the patterns.**

2. **Analyze and verify that the patterns are real.**

3. **Identify the causes and consequences of these patterns based on your knowledge of the study area.**

4. **Apply this knowledge in your own profession for prediction and planning.**

The data inside the GIS allow you to do all the steps in the preceding list except one. The display shows the patterns so that you can visualize them. The analytical techniques allow you to quantify and summarize the patterns for verification. Finally, the software allows you to create new scenarios and model the consequences of change to help you decide which of your scenarios works best. But the GIS can't tell you what the functional relationships are. It allows you to test different ideas and make pattern comparisons, but you, as the expert, have to decide which factors to examine and which patterns to compare. GIS is an excellent vehicle, but — like any vehicle — it's only as good as the person driving it.

GIS specialists can't always apply the software to effectively solve problems because they're often not subject matter experts. So, subject matter specialists need to communicate the possible causes and consequences of patterns to the GIS applications developer. For example, a change in traffic patterns might indicate possible traffic congestion. Traffic congestion might further suggest a need for new road construction, which results in even more changes in traffic patterns.

Part II
Geography Goes Digital

The 5th Wave   By Rich Tennant

UBER-USER DWAYNE GRANTZ CHALKS
UP BEFORE PUTTING ARCGIS
THROUGH ITS PACES.

In this part . . .

We live in a digital world, and that world is filled with all sorts of geographic features, surfaces, and distributions. In this part, I explain how getting the geography into digital form requires that you understand the different data models. You see how each model works and discover its impact on the way you store, retrieve, edit, and analyze geographic data. Dive into this part and get ready to feel much more comfortable with digital map data.

Chapter 4

Creating a Conceptual Model

M aps are complex devices that hold tons of information. Some of that information appears outright in the symbols, but a map imparts much information more subtly, almost covertly. A single shading pattern that represents the distribution of a soil type explicitly shows where the soil is located. But such shading patterns also tell you what soils are adjacent to other types of soils, the relative amounts of each soil type, and even how often the types occur in a single piece of land.

For another example, if your map has point symbols that represent the locations of churches, each point symbol links explicitly to a church's absolute map coordinates. And when you look at a church symbol on an area map, you can also see a measurable distance to other churches and to potential churchgoers, how many churches are in the area, and their locations within the area's topography. If you're looking for the Church on the Hill, you can probably find it on a hill.

Maps are complex, and human map readers (like you) interpret much of the information that those maps contain. So, to create a GIS that has both complex and useful information, you have to show the computer how to think like you do as a map reader.

Formulate a *conceptual model* (a picture in your mind) of how you plan to tell the computer all the information that you glean from a map. In this chapter, you can discover how to think very precisely about a map so that you can transfer that precision into your GIS projects.

Helping Computers Read Maps

A traditional map shows the properties of its features as a picture, and skilled cartographers do a good job of drawing the picture. When you see, read, and interpret the map, you often recognize properties because you know what the symbols in the picture mean. You might easily understand that topographic contour lines are close together on a map when the slope is steep, and they're farther apart when the slope is more gradual. When you see a church on a hill, you know that it's on a hill and that the people have to go up that hill to get there.

Then, along comes a computer, which really can't read a map at all on its own. To the computer, contour lines are just a graphic representation of numbers located in its files. It doesn't understand that the lines represent elevation values. Likewise, the computer doesn't understand the kind of information that you glean from the map at a glance — for example, that a certain symbol represents a church, that the church is on a hill, that the roads leading up to the church must go uphill, and that those roads connect to a network of other roads leading to the people who attend that church.

You have to tell the computer all this stuff by constructing a conceptual model. You first figure out how the map was constructed and why it was made that way. You determine what decisions went into selecting the map areas, the categories of features included, and the graphics that represent the features. With these details in place, you can make a model that even your computer will love (or, at least, understand).

Embracing the Model-Creation Process

You can't create useful GIS projects until you have a grasp of what you want to produce. In other words, you start the model-creation process with the big picture by asking what your final product needs to be. Then, you break down, or *decompose,* the final product into the general types *(themes)* of maps that you need to use. The last stage in the process involves determining which aspects of each map you need to examine. This final stage actually defines the map *elements* (features and their related data) that you ultimately plan to combine to create your final map.

Each GIS project has its own map elements, and each has its own proposed outcome. To decide what map features you need, you can just draw a simple flowchart (which is what I like to do), starting with the final outcome and successively breaking it down into themes and individual elements. I recommend that you create this flowchart without thinking about whether these data are available. That way, you devise the most detailed model with no compromises.

Defining Your Map's Contents

Before you create a map, whether it's *analog* (hard copy) or *digital* (electronic), you first need to be aware of the geography that you want to include. What elements you put into the maps, and what maps you put into the GIS, ultimately determine the effectiveness of your analysis.

Each particular subject has many kinds of geography associated with it. For example, you might be working with economic geography, medical geography, transportation geography, urban geography, and physical geography — and literally hundreds more. Each geography relates to the types of data you want to use in your own projects. So, for example, suppose you're using GIS to improve the productivity of your stores — you rely on these geographies of your area:

- **The economic geography:** Points out economic factors — such as average income of shoppers in your area — that apply to your stores

- **The transportation geography:** Gives you an understanding of the transportation systems that people use to move to and from your stores

- **The physical geography:** Helps you understand what impact climate and weather have on what you sell and when

Your mission, should you choose to accept it, is to pick the most appropriate geographies and the most useful parts of those geographies to build your GIS and its component maps.

Choosing a theme to map

I have a very simple trick that I use to decide what map or maps I'm going to include in a GIS. First, like with any project, I define the final product. If you plan to use GIS to define pieces of land that you can use for a housing development, for example, your final map should show you a list, preferably prioritized, of possible locations for your land purchase. To outline the specifications for the content of the final map, you ask yourself these basic questions:

- How big a piece of land do you need?
- How much are you willing to pay for the land?
- What are your locational constraints? Where would you prefer the land to be?
- What are the legal constraints — such as zoning and ownership — for developing your land?
- What physical constraints on the land are you willing to accept?
- What are the infrastructure needs for this development?

Your answers to the preceding questions suggest what data you need to put into the computer so that you can produce the final map you need:

- ✔ You need a map that includes all the known land parcels so that the software can measure the size of each for you and find parcels that fit within your size requirement.

- ✔ A map that includes the current sale price of the land allows the GIS to find parcels of land that are in your price range.

- ✔ If you need land near hospitals, schools, the mountains, or other features, maps for each of these features can help you measure proximity to them.

- ✔ Maps of zoning help you determine whether the land is zoned for residential development or whether you need to get a zoning variance to build there. You also want a land ownership map so that you know from whom you need to purchase the land, or even whether you can buy it at all.

- ✔ You can map topography, soils, vegetation, bedrock, and other physical factors, and each factor provides important information that you can use to decide how much land manipulation you need to perform to develop there.

- ✔ Maps that show availability of power, water, curb and gutter, and telecommunications systems help you understand if you have to pay to have these infrastructures put in.

By starting at the final product, you can break the final map down into the individual data component that are required to construct it. In the preceding example, you specify the details by asking questions about what you need to know to buy a parcel of land for development. You then translate your questions into the individual mapped features that you need to answer those questions.

Applying the methodology to any GIS project

If you don't want to purchase land, or build homes, or anything related to real estate, you may be wondering if the example in the preceding section even applies to you. What if you're trying to locate a new store, find the best place to establish a wildlife preserve, determine where crime is most likely to occur, or predict the risk of exposure to a hazardous waste cargo spill? Each scenario, no matter what the subject matter, still asks some question about location — where features are, where they could be, and/or where they should be.

Most major GIS decision-making operations deal with the locations or distributions of features. Sometimes, you use GIS to describe the impact that existing locations or distributions have, and other times, you use it to make decisions about where you want or need to put things such as businesses, public facilities, services, research facilities, personnel, or almost anything you can think of that occurs on the Earth. So, whatever your application, the methodology applies.

No matter what topic you're mapping, you can best begin by specifying what you want as the final product. When you know what your final map should consist of, you can separate and categorize its relevant parts.

Breaking down the data you want to include

Each of the maps you select to include in a GIS database is composed of preselected geographic features, and each of those features has various geographic data associated with it. In some cases, you need all the data about a particular feature, and in other cases, you need only a selection.

Consider the housing development example from the section "Choosing a theme to map," earlier in this chapter. You might find a map of the soils in your development area useful, but you really don't care about the name of the soil nor whether it conforms to the latest soil survey classification methods. You do want to know about the engineering properties of the soil, however. Can the soil support housing, or do you have to bring in soil from somewhere else? If you have to build a septic system, does the soil support that activity?

At this point in the model-creation process, you need to outline details of your project and ask the specific questions that tell you exactly what data you need from the thematic maps you've already chosen. The housing development project involves more than just land, and you need to outline all the pieces to end up with a complete database for your GIS. By creating this outline, you break the problem into neat chunks that you can more easily work with.

Say that you want to find a place to locate a bird sanctuary as part of your housing development, so you need to find a chunk of land that meets your project requirements of availability, size, price, location, accessibility, and usability. You have the following map themes at your disposal to help you find the right location:

✔ **Land ownership:** From the land ownership map, you can extract which owners you need to contact after you find out whether their land is for sale. You may not be able to buy from certain land owners, such as Federal and State government agencies. So, you want to separate out private land from public land.

✔ **Land for sale:** This map tells you which parcels of land are actually for sale, how big they are, and (hopefully) the price per acre. Your goal is to find a parcel of at least 20 acres that costs less than $2,000 per acre. And you want a single parcel of 20 acres, not a bunch of little parcels that add up to that size.

✔ **City limits:** Because you want to have people visit the sanctuary, you want it to be near (less than 5 miles from) town. To know how far your sanctuary would be from town, you need a map that shows the city limits.

✔ **Transportation:** Being close to town doesn't help you if you have no transportation system. And you want people to be able to travel on two-lane or wider roads (for example, highways). These require you have a transportation map so you can compare the locations of parcels to the locations of the roads.

✔ **Utilities:** You want to have electricity, gas, and water available at the bird sanctuary site. This requires a map of utilities so you know how far away potential land parcels are from them.

Figure 4-1 shows a flowchart of the process you go through to break down your project details and relate them to the information in your thematic maps. The final map (the bird sanctuary) is on the right, and you can see all its components on the left. Table 4-1 also shows a breakdown of project details, with the questions you ask along the way to identify the data you need to select.

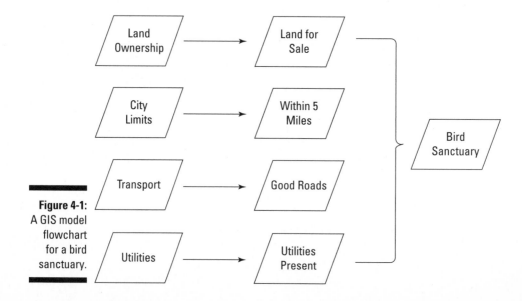

Figure 4-1:
A GIS model flowchart for a bird sanctuary.

Table 4-1	**Selecting Data from Map Themes**		
Project Detail	*Theme*	*Questions to Ask*	*Data to Select*
Land that can be purchased	Land ownership	Is the land publicly or privately owned?	Privately owned land
Land that's suitably sized and priced	Land for sale	Which parcels are for sale? How big are the parcels? How much do they cost?	Parcels for sale Size is at least 20 acres Cost is less than $2,000 per acre
Land that's appropriately located	City limits	How far are the parcels from the city limits?	Parcels within 5 miles of city limits
Land with access to adequate transportation	Transportation	What kinds of roads provide access to the parcels?	Parcels accessible by 2-lane or wider roads
Land with public utilities available	Utilities	What utilities are available for which parcels?	Parcels that have access to public water, electricity, and gas

Verifying your data's characteristics

The preceding sections show you how to select appropriate map themes and then divide those themes into the individual data components that you need to match project details and answer your questions. So, you've identified the data components, but you need to know a little more about how those components are represented in a GIS as data.

Each digital thematic map is part of a larger group of digital maps that you use to answer your GIS question. Each map has features represented by points, lines, areas, and volumes that contain useful information for you. If you're just starting your current GIS decision-making process, you need to know what data you have in the GIS database in the first place. And you also need to know the characteristics of those data, including

✔ **Level of data measurement:** The data measurement available in the GIS can be nominal, ordinal, interval, or ratio. (See Chapter 3 for an explanation of these data measurements.)

> ✔ **The level of detail:** The scale of the mapped data helps you determine the level of detail available (larger scale gives you greater detail) and how the features are represented. A feature that looks like a point on a small-scale map probably shows up as an area on a medium- or large-scale map. (Check out Chapter 2 for more on the importance of scale.)

All the questions you ask and answer to outline the GIS data you need are specific to your own project. My basic rule of thumb is to be comprehensive with your questions so that you can select the best, most complete set of data possible. You can generalize from detailed data much more easily than you can get specifics from general data.

Converting from Map to Computer

Computers can help you map your study area and analyze those maps, but only if you plan carefully what data you include and how you input that data into the computer. When you select your map's content, you include the specific point, line, area, and surface features that can answer your project questions. But your computer doesn't know as much as you do about maps.

In GIS terms, *polygon* refers to area, and *surface* means that a feature has a third dimension (for example, height).

Deciding how to represent your map

Cartographers use two general data forms that translate the map into digital form: *raster* (little squares or grid cells) and *vector* (or points, lines, and polygons). So raster data resemble a chessboard, and vector data resemble line drawings. Most GIS software today can handle both types of data, so don't panic about which GIS software you need based on the type of data you want to use. Each conceptual model base (raster or vector) brings with it a unique set of tools, levels of accuracy, ways of modeling, and advantages and disadvantages, as covered in the following sections.

Picking a raster-based model

Understanding how raster-based GIS software is constructed and how the databases work with it is really pretty easy. This method will probably make perfect sense to you when you conceptualize your raster model. Before you know it, you're thinking like a chess master, seeing squares everywhere and relating them to your next GIS move.

Think of raster, or grid cell, GIS like a checkerboard or chessboard. Each square is a single grid cell, and each grid cell represents a portion of the Earth. When mapping, you assume that portions of the Earth are square for simplicity, and the GIS software can adjust the Earth coordinates for this assumption.

Raster GIS represents all forms of geographic features by using square grid cells, each of which represents a portion of your map. You can conceptualize each of these grid cells as a part of a checkerboard. The following bullets take you through the thought process for each feature type:

- **Points:** In a conceptual model of geography, points have no dimensions (length or width). Of course, you have to take liberties to use grid cells to represent dimensionless points because grid cells do have both length and width. If you make these grid cells really, really small, like the dots on a TV screen, this whole idea is a bit easier to accept.

  So, if you think about raster GIS as a checkerboard, a single grid cell is a point object (see Figure 4-2), such as a church, mosque, or synagogue. Each different type of place of worship, such as churches, has a unique grid cell color on the raster map that's associated with a different number inside the computer. So, unlike a checkerboard, you can have more than two colors of squares. Say that you use the number 1 to represent churches, 2 to represent mosques, and 3 to represent synagogues. These different types of features all have something in common — each represents a place of worship associated with a common theme or map layer. This common characteristic limits the number of different codes and colors that you might need for your map of point features.

- **Lines:** In grid cell–based GIS, a string of grid cells represents a linear object (such as a road, a railroad, or a walking path). This string of cells can be lined up *orthogonally* (edge to edge), *diagonally* (along the corners), or some combination, depending on how curvy the feature is. Line objects are considered to have only one dimension (length), even though they usually have some width in the real world. So, when you use grid cells to represent lines, you have a bit of the same problem you have with point features — you have to use a symbol that has area to represent an object that doesn't.

  Just like with point features, on a raster map you use different colors or patterns for each line category (see Figure 4-3), and the computer associates those colors or patterns with particular numbers. So, when you input these, you might have the computer program assign roads a value of 1, railroads a value of 2, and walking paths a value of 3. All three of these features share a common theme — in this case, transportation.

✔ **Areas (polygons):** The grid-cell system can easily represent areas. The computer symbolizes each unique area — for example, crop types such as corn, soybeans, and potatoes — as a single grid cell (if the area is small enough) or a group of grid cells, as shown in Figure 4-4. And these cells may be connected *(contiguous)* or disconnected *(non-contiguous)*. The size of each grid cell corresponds to a certain area of land. So, if a single grid cell corresponds to a land chunk that measures 10 meters on a side, that cell represents 100 square meters (10×10 meters) of land area.

Like with point and line features, each crop type shows up on a raster map as grid cells that have a common color, shade, or pattern, and each type has a unique number associated with it inside the computer. In the crop example, you might use a value of 1 to represent corn, 2 for soybeans, and 3 for potatoes.

✔ **Surfaces (volumes):** When you're representing surfaces or their volumes, each grid cell has, in addition to length and width, a number associated with the height or depth of the space. This number may represent elevation above sea level or depth to groundwater for a point at the center of a grid cell. In some GIS, grid-cell values even represent nonphysical surfaces, such as population density or land appraisals.

Grid cells that represent surfaces have so many possible values that each grid cell may have its own unique number. To go along with this number, on a raster map, you could give each grid cell an equally unique color or shading pattern. But if you do, you soon find that — with so many different looks — you can't really distinguish one from the other.

To make the task of assigning numbers a bit easier, the GIS software usually allows you to designate surface values by groups. For example, if you want to represent elevation, you might have one group of grid cells that ranges in elevation from 0 to 1,000 feet, another group that ranges from 1,001 to 1,500 feet, yet another group that ranges from 1,501 to 2,000 feet, and so on. When you display your grid cells that represent the range of values (like in Figure 4-5), you can actually distinguish among the various elevations. You have control over how you group values to represent surfaces, but the computer still keeps all the myriad original unique values intact.

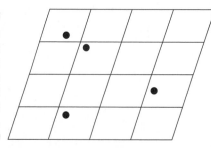 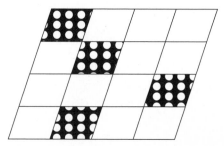

Figure 4-2:
The point features on the left are represented as grid cells on the right.

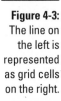

Figure 4-3:
The line on the left is represented as grid cells on the right.

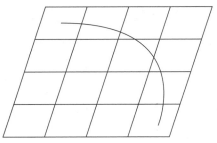 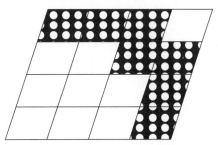

Figure 4-4:
The area feature on the left is represented as grid cells on the right.

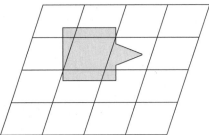 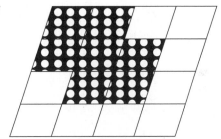

Figure 4-5:
Surface features on the left are represented as grid cells on the right.

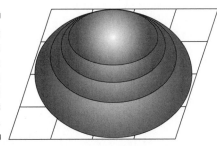 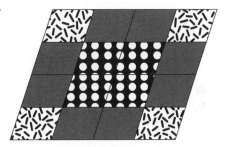

Raster GIS represents points by using a single grid cell, lines by using a line of grid cells, areas by using a group of grid cells, and surfaces by using groups of grid cells that have additional unique values.

Choosing a vector-based model

In a vector-based conceptual model, representing point, line, and area features is pretty straightforward. The vector data structure represents each point object by a single X and Y coordinate pair, each line by a set of two or more X and Y coordinate pairs, and each area by at least three lines that create a closed figure (the lines' start and end points coincide to create *closure,* as shown in Figure 4-6).

Figure 4-6:
The
vector data
structure, as
represented
by X and Y
coordinate
pairs.

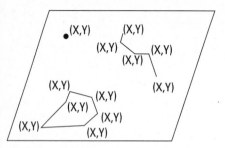

Representing surfaces or volumes by using vector data is a bit more difficult than representing points, lines, and areas, but the GIS industry has a well-established and commonly used data structure. This data structure, the *Triangulated Irregular Network (TIN)*, represents surfaces and volumes as a collection of non-overlapping triangles, which is very similar to the way that computer games model 3-D objects. The *vertices* (points) of each triangle have unique X, Y, and Z (height) values, but each group of three vertices creates a triangular plane whose surface has a unique angle and direction (as shown in Figure 4-7).

The neat thing about the TIN model is that you can use it to predict *(interpolate)* missing values, create cross sections through surfaces and volumes, draw contour lines, and create 3-D visualizations, just like its raster equivalent data model (see Figure 4-5).

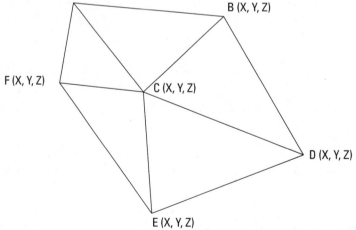

Figure 4-7:
The
Triangulated
Irregular
Network
(TIN) model
of non-
overlapping
triangles.

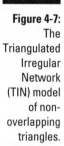

Weighing the benefits: Raster versus vector

You may wonder how you should decide whether you need raster or vector data (described in the preceding sections) — or both — in your GIS operation. The raster-versus-vector debate is far less important today than it was in the past. Most modern GIS software has both raster and vector data components, or it can convert from one format to another. You need to determine the most important functionality, accuracy, and storage issues for the work you want to do when you decide on GIS software.

So, to help you make your decision, here's my nutshell look at the differences between raster and vector data models:

- ✔ **Raster data takes up more computer space than vector data.** Space may not be a serious issue unless you know that you have limited computer storage and can't get any more. Raster data require that each grid cell have an associated value, but as few as three sets of X and Y coordinate pairs can represent large areas in vector data.

- ✔ **Raster is faster (computationally) than vector.** With computer speed rapidly increasing, this speed difference might not seem a huge issue, but the spatial datasets that GIS use are increasing in size at an even faster pace. So, if you have GIS applications that involve many datasets, you may want to use the computationally faster raster data.

- ✔ **Raster is compatible with much scanned and remotely sensed data.** Many raster datasets, usually in the form of scanned images (aerial photography) or satellite images, are quite useful in GIS analysis. Because they are similar data structures, you don't need to do a lot of data conversion to add them to your GIS. Determine what datasets your application needs and keep those datasets in mind when you choose your data structure.

- ✔ **Raster is less spatially accurate than vector.** A lot of people believe that vector data gives an accurate representation of locations in geographic space, but that belief isn't entirely true. Computers can't represent geographic space exactly because each computer's internal computational accuracies vary. Computers often truncate or round up numbers, have single or double *precision* (the number of points beyond the decimal point that they round to), and manipulate numbers by using algorithms that do strange things (other than rounding) to the numbers. Even so, the spatial accuracy of vector is much better than that of raster, where the *resolution* (spatial dimension) of the grid cells determines accuracy. The smaller the grid-cell size, the more accurately it represents geographic space.

✔ **Raster data give you more options to work with surfaces than vector data.** Because each grid cell has a unique surface value, any measurement made over the entire surface can reflect these unique values, whereas with vector data (the TIN model), the computer assumes that large parts of the surface have the same value (the triangular plane's slope).

✔ **Raster generally provides visually less desirable output (especially with its coarser resolution) than vector.** Some map readers dislike maps represented as grid cells because the lines and edges do not appear smooth like they would in paper maps.

✔ **Raster is more powerful for modeling than vector.** One other factor, and an important one for the modeler, is that you can use raster data much more efficiently as data for modeling because it gives you far more options. In Chapter 18, you can examine the detail of the Map Algebra modeling language that was originally developed for the raster data model. You can use the language to perform complex processes, called *cartographic modeling,* with more than just raster data, but the language has far more options with raster data. The raster data model has nearly limitless and extremely powerful (complex and accurate) modeling capabilities.

✔ **Raster is more compatible with printer output technology and less compatible with plotter output technology than vector.** (A *plotter* is a type of printer that draws lines rather than printing a series of dots.) Raster and vector models in the GIS software must communicate with hardware for input and output. Each input and output device also has its own coordinate system and its own data structure (whether raster or vector).

You can convert the data captured by most input devices — whether vector (as in the case of digitizers) or raster (as in the case of scanners) — from one to the other data model without much loss of accuracy. On the other hand, plotters tend to be vector based, and printers tend to be raster based. The compatibility of your GIS data model with that of your input and output devices becomes a factor in determining the quality of your output.

Check out Chapter 22 for guidance on selecting a GIS vendor.

Chapter 5

Understanding the GIS Data Models

· ·

· ·

*W*hen you read a map, you must be able to visualize what the real world represented by the map looks like. As a GIS analyst, you need an understanding that goes beyond looks because you use this symbolic information to analyze and combine data, and more importantly, make decisions about the space that the map represents. In some cases, you lose the graphical elegance of a paper map in the computer version because the computer emphasizes computational effectiveness over good looks.

In this chapter, I help you understand how the data models that the software uses work so that you understand both the limitations and the power of a GIS.

Examining Raster Models and Structure

A checkerboard represents a unit of space divided into squares upon which the checkers rest. Likewise, raster GIS grids (the checkerboard) are broken up into grid cells (the squares) upon which geographic features (plants, animals, houses, towns, rivers, and so on) rest. On a checkerboard, the checkers are like occupants of the geographic space — you can move them around in different places. But the checkers have to stay on the dark-colored squares and move between them diagonally. In other words, the checkerboard restricts where you can place features (checkers) and where you can move them. Similarly, geographic space has limitations about what can and can't occur.

In the real world, the nature of physical features imposes rules that control or limit what things can or should occupy different spaces. For example, you can't build houses in the middle of a river, and you can't sail boats on dry land. But the real world doesn't consist of alternating squares of water and land; the makeup of the Earth is much more complex than that. Figure 5-1 shows a grid overlaid on a map that depicts a more typical distribution of land and water features.

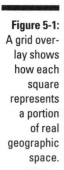

Figure 5-1:
A grid overlay shows how each square represents a portion of real geographic space.

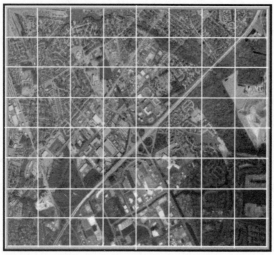

Representing dimension when everything is square

If geographic space were exactly like a checkerboard, it would be composed of square areas, each containing a given category of Earth surface data. But real geographic space contains more than area (or polygon) features. Real geography has point, line, and surface (or volume) features that must also show up in the grid structure.

Surfaces require that the grid system represent a third dimension (and a bit more complexity); I discuss how the computer represents surfaces in the section "Dealing with Surfaces," later in this chapter. In this list, I show you how to represent features that have two dimensions or less by using grid cells (see Figure 5-2):

✔ **Points:** Even though points are considered zero-dimensional objects that don't take up any space, they're still included in the raster representation as a single grid cell.

✔ **Lines:** Features represented as lines (one-dimensional objects) are composed of a string of grid cells.

✔ **Areas:** By definition, areas are two-dimensional objects, and they show up as groups of grid cells.

In raster GIS, points are single grid cells, lines are strings of grid cells, and areas are groups of grid cells. See Chapter 4 for a more complete discussion of how grids represent geographic features.

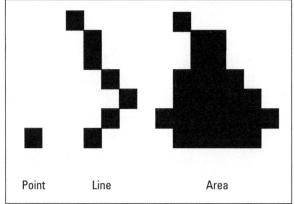

Figure 5-2:
Point, line,
and area
features in
the raster
data model.

Point   Line   Area

Making a quality difference with resolution

At first glance, representing zero- and one-dimensional objects with two-dimensional grid cells doesn't seem to make much sense. But the raster model is just a way of representing geographic space and its features to help you store related information in the computer.

The spatial accuracy does suffer with these grid cells, especially for points and lines. This data model compromises both the exact location and the amount of space occupied. But this limitation in spatial accuracy is the necessary trade-off with the power of the raster data model for performing high-quality geographic modeling. Given this trade-off, you have one good way to improve the spatial quality of grid cells — make them smaller. Like with a high-definition television, the smaller the dots, the higher the resolution, and the more accurate the representation. In raster GIS, this concept is called the *grid cell resolution.*

In raster GIS, the smaller the grid cell size, the finer its resolution, and the more accurately those cells represent the location and spatial extent of the geographic features.

Finding objects by coordinates

The raster coordinate system is a set of sequential numbers that identify the column (X) and row (Y) locations of each grid cell. The sequence usually starts with the upper-left corner of the grid and moves from left to right across the columns and from top to bottom down the rows (see Figure 5-3). This X and Y coordinate system is pretty standard, but it doesn't preclude starting, for example, on the bottom left and counting up and to the right. In any case, all grid coordinate systems do basically the same thing — they locate the positions of grid cells based on columns and rows.

Because the grid-cell resolution represents real geographic space, any movement from one cell to another constitutes a symbolic traverse of both distance and direction in geographic space. For example, if you set your grid-cell resolution at 100 meters, each grid cell is 100 meters on a side and has an area that you can calculate as follows:

$$100 \times 100 = 10,000 \text{ square meters}$$

So, if you start at the upper-left corner of a map represented this way and move 15 grid cells to the right, you're in grid column 15 and grid row 1. It also means that you traveled 15 times the length of each grid cell (1,500 meters). And you can link this distance and direction directly to the geographic grid (latitude and longitude), which means that you traveled 1,500 meters east from your starting point. Likewise, if you move down a column, you can envision that you're moving south.

This system allows you to find virtually any object by its column and row coordinates. And, because these column and row coordinates are linked to the geographic grid, you can easily relate the grid cells to specific positions on the Earth. More important for the GIS modeler, you can layer grids and relate the corresponding grid cells to each other. Using layered grids for modeling requires that

✔ All the grids represent the same portion of the Earth.

✔ Grids are *co-registered* (meaning they lie directly on top of one another).

✔ Each grid cell is the same size in every map layer.

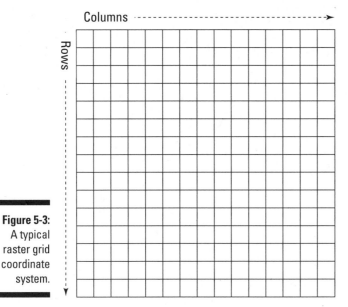

Figure 5-3:
A typical
raster grid
coordinate
system.

The really cool thing about having these grid cells co-registered is that you can compare lots of different features that occur in the same geographic location but appear on different map layers. For example, you can compare the location of a building (on the land-use layer) with its soil (on the soil-type layer) and the existence of a hazardous chemical found in that soil (on the soil-chemistry layer), all because the locations of each set of grid cells are right on top of each other in the various map layers.

Finding grid cells by category

Each grid cell in a raster GIS has a unique set of X and Y coordinates, and each group of cells has different properties. Unlike the squares on a checkerboard, grid cells come in as many types as you have categories to include in your map.

Because the raster data model structures grid cells into a coordinate system, you can find any grid cell you want by simply determining its column and row location (X and Y coordinates). Because you can also assign a category (a particular type, class, or even value) to each grid cell, line of grid cells, or group of grid cells, you give yourself another way of finding them. That is, after you make the assignments, you can search the grid by categories, as well as by coordinates.

Raster GIS gives you the power to search the grid in two ways. You can search by coordinates to examine what the grid cell represents, or you can search by grid-cell quality to find out where grid cells with that quality are located.

Working with map layers

When raster GIS was first developing, software programmers made different attempts to create a storage and recordkeeping system that would allow the user to input, store, retrieve, manipulate, compare, and output grids. Perhaps the most successful, called the Map Analysis Package (MAP), presents maps as a series of layers, each with a different name. First, you find the map layer that you need. You retrieve map layers by searching for the map name (the *theme*). Each map has a set of categories, each category with a unique number value that can be retrieved individually. These categories also have labels associated with them for display. Figure 5-4 illustrates the MAP raster data model.

The MAP data model breaks down into these components (from the most general to most specific):

- ✔ **Theme:** The major unit of retrieval. In Figure 5-4, the map themes are called GRID LAYER 1, GRID LAYER 2, and so on.

- ✔ **Category:** After you retrieve a map, you're free to search the multiple categories within the map's theme. This data model represents each of these descriptive attributes with a unique numerical value (for example, Category1 in Figure 5-4). Modern raster GIS software allows you to search these categories by name.

- ✔ **Value:** Numbers that represent the categories. You use those numbers to do your analysis. In other words, the computer actually retrieves, compares, and analyzes the numbers themselves rather than the categories they represent. A GIS locates the category values associated with sets of points (grid cells) by their X and Y coordinates. Having category values associated with grid-cell locations is essential for doing overlays (Chapter 16) and for using the modeling language called map algebra, which I describe in Chapter 17.

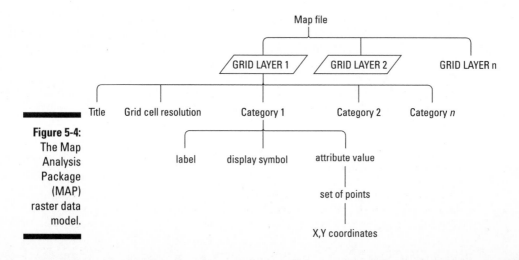

Figure 5-4:
The Map Analysis Package (MAP) raster data model.

Linking objects and descriptions

The Map Analysis Package (MAP) data model makes it easy to find individual categories in a single thematic grid. Its limitation is that, as originally designed, the MAP data model represents each category only once per layer. The solution is simplicity itself. For each map, you add not just categories, but also a database of categories by linking the map to a database management system. The database management system can store multiple values for each category. So, now you can have a map of land use that includes a category called housing, and that category can have many different subcategories. Raster software such as the Spatial Analyst that the Environmental Systems Research Institute (ESRI) uses currently implements this model.

Extending the raster data model by including a database management system gives you a lot more flexibility. But you need to make sure that you know where to look for the categories and that you give them names you can remember. Always use meaningful and memorable names for categories, ID-codes (codes you use for categories), and maps, if at all possible.

Exploring Vector Representation

To think about maps inside the computer in a graphically familiar way (think paper map), envision each point as a point with its own X and Y coordinates, each line with an ordered set of these coordinate pairs, and each polygon with a string of points that close (begin and end at the same place). Conceptually, vector GIS software both stores and displays points and lines more accurately, making those points and lines nearly zero- and one-dimensional, respectively, and having polygons more closely represent their own shape. The closer together the points that make up line segments, the more accurately they represent the real lines.

Several data models allow the conversion, storage, and manipulation of map data by using the vector approach. Some of these models are simple and work very well for input and output devices, and others are more complex and allow for some serious computational power.

Simple forms of vector representation

Simple forms of vector representation focus more on the accurate graphic depiction of features and less on the subsequent analysis of geographic information. The simple forms still in use today are mostly for input and output, rather than actual GIS functionality. Modern GIS software communicates quite effectively with these feature-depicting models, which often come built-in to the GIS as graphics languages.

This ability to communicate with graphics depiction models is necessary because even the most complex data modeling can't happen if you can't get data into the GIS software or create a map when you're done with your analysis.

Spaghetti representation

The simplest of vector data models is the *spaghetti model* (see Figure 5-5), a one-to-one translation of graphics into the computer without regard for spatial relationships. Software using the spaghetti model stores points in the computer with identifying numbers (for example, the first point is number 1, the second is number 2, and so on). Each number is linked directly to its corresponding X and Y coordinates in a separate data table. Lines are ordered strings of points (the spaghetti), each associated with a searchable point identification number. Finally, the software stores the polygons, which also have searchable IDs, as closed loops of coordinate pairs.

The spaghetti model is simple, easy to understand, and (more importantly) fast for both input and output. It doesn't work — meaning it doesn't store information correctly — when you try to represent features such as islands because they're effectively a polygon within a polygon. The data model also doesn't recognize polygons comprised of non-contiguous parts, such as island chains. These data representation limitations make this data model ineffective for modeling and analysis.

Alternative representation

One vector data model created by ESRI gives much more control than the spaghetti model. Instead of limiting storage of points, lines, and polygons to coordinate pairs, this model (called a *shapefile*) stores the geometry of each feature as a shape that contains the coordinates and links to the attributes. GIS software packages widely use this data model today because of its relatively low processing overhead, low storage requirements, fast drawing speeds, and its ability to handle overlaps and non-contiguous features. You can also, with relative ease, convert the shapefile model to more complex data models, which I describe in the section "Complex forms of Vector Representation," later in this chapter.

The term *shapefile* is a bit misleading. A shapefile isn't a single file. Instead, it's three separate files that allow for the representation of 14 different types of geometric shapes (as outlined in Figure 5-6). The primary file with an `.shp` extension is a list of X and Y coordinates that define objects called *shapes*. To keep track of the shapes and search them by type, it also has an index file that has a file extension `.shx`. Finally, the `.dbf` (or database) file contains all the attributes that describe the features.

Although the shapefile is still a fairly simple data model, compared to some of the more advanced ones, it's likely to be a GIS software mainstay data model for years to come because it's simple and compatible with many software programs.

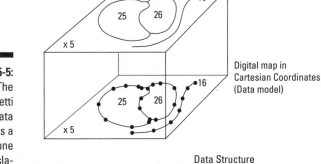

Figure 5-5:
The
spaghetti
vector data
model is a
one-to-one
transla-
tion of the
graphics
into the
computer.

Data Structure

Feature	Number	Location
Point	5	x,y (single pair)
Line	16	(string of x,y coordinate pairs)
Polygon	25	(closed loop of x,y coordinate pairs where first and last pair are the same)
	26	(closed loop sharing coordinates with adjacent polygons to form a data structure)

Value	Shape type
0	Null shape
1	Point
3	PolyLine
5	Polygon
8	MultiPoint
11	PointZ
13	PolyLineZ
15	PolygonZ
18	MultiPointZ
21	PointM
23	PolyLineM
25	PolygonM
28	MultiPointM
31	MultiPatch

Figure 5-6:
The shape-
file data
model sup-
ports these
geometric
shapes,
among
others.

Complex forms of vector representation

Despite its robustness and elegance, the shapefile (see the preceding section) lacks one thing that many high-end modelers demand — the ability to
see spatial relationships. This characteristic, called *topology* (which I define
more specifically in the next section), allows the GIS software to recognize

and convey some basic spatial relationships that humans express by using an English language construct called the *prepositional phrase.* (Don't worry. You don't have to diagram sentences.)

In everyday language, you use many prepositional phrases that describe the spatial relationships you encounter. For example, you may live "outside a town," "next to your neighbor," and "within the county." You travel roads that go "through forests," "along rail lines," and "over hills." The data models discussed in the section "Simple forms of vector representation," earlier in this chapter, don't have these relationships (known as topological relationships) explicitly coded. You need to incorporate the mathematics of these relationships — the topology — into the data model so that you don't have to calculate these spatial relationships each time you want to analyze something.

Using the mathematics of neighborhood

When you look at a map, you can easily see where everything represented on the map is relative to everything else on that map. For example, when you look at a map of your own neighborhood, you get an instant view of that neighborhood. You can see shortcuts, recognize which places are close to your house and which are far from your school, determine what's on each side of your street, and know which streets are connected to each other and which aren't. To replicate this powerful property of the human brain inside the computer, computer scientists employ a branch of mathematics called *topology* that studies the positions and relative locations of geometric figures. I like to call it the "mathematics of neighborhood" because it allows the computer to understand how a neighborhood works (including what features are located there, where they are relative to each other, and how to find them).

Because computers can't reason or look around at their neighborhoods, you have to tell them everything. And so you explain

- ✔ **Each line you create is linked to other lines.** If the computer understands these connections, you can move from one link to another, just like you'd drive your car from one street to another.

- ✔ **The classifications of nearby areas.** Say that on the left side of a line (perhaps your street) is a polygon (a parcel of land) that's agricultural, and on your right side is a single-family residential parcel.

The U.S. Bureau of the Census needed computers that understood the connections and classifications used in its work. This need led the bureau to develop two of the more well-known topological data models — the Dual Independent Map Encoding (DIME) and subsequent Topologically Integrated Geographical Encoding and Referencing (TIGER) models (discussed in the following section).

Figure 5-7 shows a generic topological vector model. All the others, including commercial ones, are based on this model, with minor variations. The left side of the figure shows several points, lines, and polygons, all of which are numbered:

✔ **Points:** The numbers that have circles around them

✔ **Lines:** The numbers that interrupt the lines

✔ **Polygons:** The numbers inside the polygons

These numbers are essential because they allow the computer to know which part you're referring to when you want to compare the relative locations of all the points, lines, and polygons.

The table at the top of Figure 5-7 contains information about the spatial relationships (the topology) of the points, lines (links), and polygons:

✔ **Link #:** Lists all the links in the figure, numbered 1 through 11.

✔ **Right and left polygons:** Tell you how the links are related to the polygons on either side of them. This is the *contiguity property* of topology, which identifies the polygon on either side of any given line. By default, it also tells you the direction of the line because it knows which polygon is on the left and which polygon is on the right of that line.

By default, polygon 0 is the outside polygon — the polygon that constitutes the area outside of the map.

✔ **Nodes 1 and 2:** Show you which *nodes* (points at the end of each link) define the beginning and ending of each link. These nodes link links to other links (whew!). So, this information tells you another aspect of topology called *connectivity* — the order and sequence of links.

Because the nodes define the links, and the beginning and ending nodes of some links occur at the same place, these links create polygons. So, if you know the coordinates of all the nodes, you also know the third basic component of topology — area. *Area,* as one of the basic properties of this topological data model, simply means that the polygons are defined by identifiable links, the coordinates of which are stored in the computer. You have that necessary information in the bottom table in Figure 5-7, which shows all the X and Y coordinates of all the nodes for all the links.

Topology is a way of coding the geometry of spatial relationships inside the computer, and topology consists of three things:

✔ **Contiguity:** The topological property that says links have a direction, as well as polygons on their left and right sides.

✔ **Connectivity:** The topological property that says nodes connect links in a definable sequence.

✔ **Area:** The topological property that says areas (polygons) are defined by identifiable links.

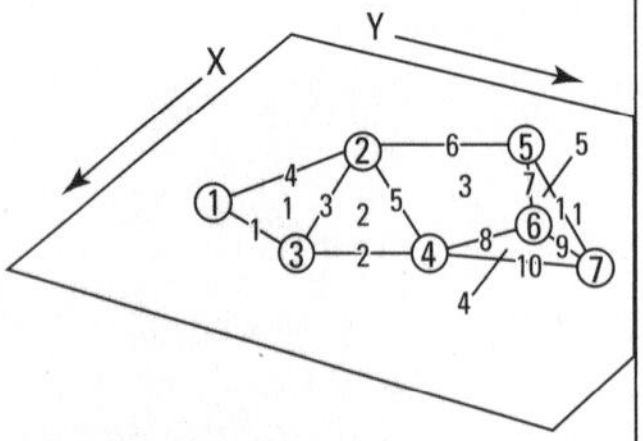

Topologically Coded Network and Polygon File

Link #	Right polygon	Left polygon	Node 1	Node 2
1	1	0	3	1
2	2	0	4	3
3	2	1	3	2
4	1	0	1	2
5	3	2	4	2
6	3	0	2	5
7	5	3	5	6
8	4	3	6	4
9	5	4	7	6
10	4	0	7	4
1	0	5	5	7

X,Y Coordinate Node File

Node #	X Coordiante	Y Coordiante
1	19	6
2	15	15
3	27	13
4	24	19
5	6	24
6	20	28
7	22	36

Figure 5-7:
A generic topological data model.

Using the U.S. Census file types (DIME and TIGER)

The United States Bureau of the Census wanted to produce maps of its street, address, and associated census data for the 1980 centennial census. Specifically, it needed to know what was on each side of a street, and to do so, needed to incorporate topology. So, the Bureau created a topological data model that supported the census geography provided by data collected every ten years — the Dual Independent Map Encoding (DIME) file. DIME was one of the first really useful datasets that was available to the general public for GIS. It also greatly enhanced the ability of researchers to examine large segments of the population through what's now called *census geography*.

DIME recognizes the three basic components of topology (contiguity, connectivity, and area). The primary purpose of this model is to allow census data users to link the tabulated census information to census geographic units (such as streets, blocks, districts, and so on). The DIME model stores the topology by creating explicit From and To nodes that define each link for each map (as shown in Figure 5-8). Each of these links, like any topological model, includes explicit information about what's on its left and right sides.

In Figure 5-8, the left and right sides of the link from node 21 to node 22 are census blocks 88 and 90; these blocks constitute polygons on a map showing these census divisions. DIME files also include addresses for each side of each link (left and right blocks) so that the DIME model also allows the user to easily identify and locate census divisions inside the database.

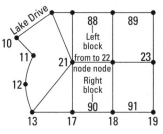

Figure 5-8:
The DIME
topological
data model.

From Node		To Node		Blocks		Left Addr		Right Addr	
Map	Node	Map	Node	Left	Right	Low	High	Low	High
3	21	3	22	88	90	111	1999	102	189

The major drawback to the DIME file is that its geography isn't very realistic. It represents a curved street, such as Lake Drive in Figure 5-8, with a straight line. In 1980, the Census Bureau wasn't concerned about the accuracy of such features. When others began using the DIME files for implementations other than census tabulation, the geography became more important and was added to the more advanced TIGER data structure. For example, some insurance companies use the census files to map which houses lie within zones likely to be flooded every 100 years (100-year flood zones). The improved geographic accuracy of the TIGER files is vital for accurately locating such houses.

The DIME files are still available today, although the U.S. Census Bureau has replaced them with a more advanced topological structure called Topologically Integrated Geographical Encoding and Referencing (TIGER). Advantages of TIGER files over DIME files include:

✔ **Easy retrieval:** The TIGER model directly addresses points, lines, and polygons, so you can more easily retrieve a census block by retrieving the block number, rather than having to find the links first and use their topological information for retrieval. Figure 5-9 illustrates how the TIGER data model is structured.

✔ **Improved geography:** Although Lake Drive is represented as a straight line in DIME files (as shown in Figure 5-8), TIGER files show the actual shape of Lake Drive (see Figure 5-9). This feature makes TIGER files much more compatible with non-census-related research and other data files that are more geographically accurate.

TIGER files are readily available both from the U.S. Bureau of the Census and third-party vendors who often repackage the files. TIGER files are compatible with a vast array of other data types, possess lots of data already in this format, and constitute an elegant implementation of the topological data model, so they're likely to be around for a while.

Understanding the ESRI coverage approach

One of the early successful topological data models, the *coverage* model, developed into one of the most successful commercial GIS products of its time — the ArcInfo that ESRI uses. Each map layer is a single unit composed of geographic features associated with a common theme. Each theme covers the study area's geographic extent, and a series of these themes represents a more comprehensive view of the content of the study area.

In a coverage, here's how the features are stored:

- ✔ **Entities:** Primary features such as points, arcs (the word ESRI uses for lines), and polygons
- ✔ **Complementary information:** Secondary features such as
  - **Tics:** Points that show the input device where the map coordinates are
  - **Links:** Computer code that ties the graphics to the descriptions
  - **Annotations:** The text of the descriptive information

The attribute information itself is stored in a separate set of database files.

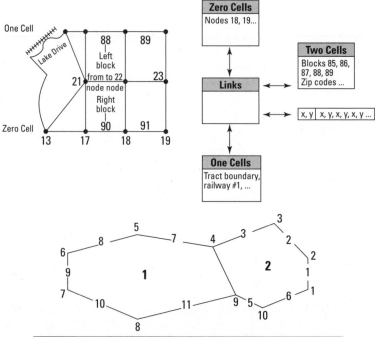

Figure 5-9: The TIGER topological data model.

Polygons		Chain list	Chains						
Polygon name	Polygon pointer	Chain list	Chain name	Chain points (X, Y strings)	Chain length	From node	To node	Left polygon	Right polygon
1	–	4	4	x, y, x, y	–	4	9	2	1
2	–	5	5	–	–	–	–	–	–
–	–	6	6	–	–	–	–	–	–
–	–	–	–	–	–	–	–	–	–
–	–	–	–	–	–	–	–	–	–

The coverage model supports the three major properties of topology: contiguity, connectivity, and area definition. The coverage model has been around for over two and a half decades, but most GIS software vendors still support it because of the large number of GIS users whose data were (or still are, in some cases) stored as coverages.

Working with object-oriented representations

Advances in computer languages and their associated data structures — the development of object-oriented programming (OOP) — brought about a useful adoption for GIS data models. OOP focuses on objects where an *object* is a set of computer code that can be copied from place to place in a computer program. In GIS, an object represents a type of geographic feature that you can move around and whose properties follow it (are inherited) no matter where you place it. Incorporating OOP languages resulted in wholesale rewriting of some GIS software and development of new software based entirely on this model.

Much of the power of OOP data models results from the objects' shared properties, or the characteristics they have in common with other objects in their group. (I describe the sharing of properties briefly in the sidebar, "Sharing the property," and talk more about object structure in Chapter 6.) Because of shared properties, an object always knows what it can and can't do, and what it can and can't support. So, you don't have to worry about any surprises when you retrieve information from an object-oriented GIS. If you're looking for information on a land parcel, you won't pull data related to oceans because the ocean objects don't share the properties of (and, so, aren't grouped with) land parcel objects.

You can find several implementations of the object-oriented model available, including the ESRI Map Objects product and the GE Smallworld GIS product.

Taking on advanced geographic representations

Perhaps the most robust and advanced of the object-oriented models is the ESRI new geodatabase model. The *geodatabase* (not to be confused with a collection of geocoded data inside a GIS software package that's sometimes referred to as a geodatabase) allows you to store a wide variety of data types (including raster, vector, CAD, and others) in the same system. Here's a quick list of the types of data this model can incorporate:

- ✔ **Attribute tables** from typical RDBMS (see Chapter 6).
- ✔ **Geographic features** that you might find in land-use maps.
- ✔ **Satellite and other digital imagery** such as the files you might get from the LANDSAT satellite.
- ✔ **Surface models** you could obtain from the USGS Digital Elevation Models (DEMs).
- ✔ **Survey data** from scientific vegetation samples, or telephone surveys, for example.

Sharing the property

In object-oriented models, geographic features are stored as objects. There's debate about exactly what an object is, but in GIS, an individual object (with its local properties) also represents a group of objects that has a set of more global properties. By being part of the group, the object shares those global properties. So, for example, a land parcel object has unique identifiers — owners, value, size, shape, and so on — but it still possess a set of properties shared by a larger group of polygons called parcels. The parcels in this group share a set of conditions that differ only in the specifics. For example, all the parcel objects have a type of land use, an owner, and zoning restrictions. So, the general properties outlined for a parcel object (meaning that it has a land-use type, an owner name, and a set of zoning restrictions) are inherited by every other parcel.

The geodatabase model, like the non-object-oriented coverage model, has tables of data that include *objects* (rows in the database table) and *features* (objects that have explicitly defined geometry). It's a completely topological model, like its predecessors. What makes the geodatabase so powerful is that each of these objects can contain a wide array of rules and logic, which makes the objects more like real geographic objects. For example, if you have a geodatabase of electrical infrastructure, certain lines that represent a type of wiring don't connect to non-compatible types of wiring because the database explicitly stores properties that prohibit such a dangerous connection. Or, if you have a road-network geodatabase, the rules might prohibit you from connecting a one-way street going east to another one-way street going west.

Although the geodatabase is new and quite different from its non-object-oriented cousins, it has a lot of tools that you can use to migrate your data to the geodatabase format. While GIS analysis becomes both more sophisticated and more realistic, the geodatabase and other object-oriented models like it will increase their capacity to represent rules and conditions that characterize real-world geographic features.

 ESRI has made many industry-specific geodatabase models that contain complete logic available for use, so you don't need to build them from scratch. If you're looking for a GIS for a specific application, check to see whether ESRI has the model you need before reinventing the wheel.

Dealing with Surfaces

Surfaces of all kinds are very common in GIS, but because of their three-dimensional nature, they place additional demands on the computer programmer who wants to represent them. Beyond having to represent the X and Y coordinate, and the associated attributes, you also have to worry about how to effectively include the third dimension in such a way that the software can perform all the basic operations on surfaces that the other data types allow, such as storing, retrieving, editing, analyzing, comparing, and outputting.

Storing surface data in a raster model

The general raster model provides the easiest method of storing surface data. Each grid cell represents not just an area on the surface of the Earth, but also a single elevation value. So, a grid cell in column X and row Y would store a single value Z that represents the elevation for the entire grid cell (as shown in Figure 5-10).

This model does limit the accuracy of your elevation values, just like the raster model itself limits the locational accuracy and spatial representation of the actual geography (see "Making a quality difference with resolution," earlier in the chapter). On the plus side, the model does allow for some very nice surface modeling with low computational overhead (far fewer calculations), and you can easily convert it to other forms of surface representation.

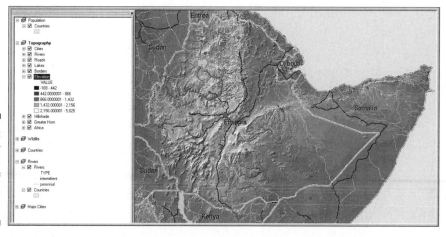

Figure 5-10: A raster representation of surface data.

The most common form of raster surface model is called the Digital Elevation Model (DEM), which was developed by the United States Geological Survey (USGS) to store and eventually process topographical surface data. DEMs are a series of regularly placed X and Y coordinate locations that also contain Z values, which depict the topographic elevations at those locations.

DEMs come in a number of different sizes, based on the maps from which they were derived. USGS DEMs come in 7.5-minute (meaning 7.5 minutes of latitude by 7.5 minutes of longitude each), 15-minute, 30-minute, and 1-degree versions. The 7.5-minute series provides the most elevational detail. When you move from 7.5 minutes to 1 degree, you lose both horizontal and vertical accuracy.

You can get the DEM directly from the USGS, and it's also available from other vendors as a value-added product. Most modern GIS packages support the DEM data model, which allows you to convert the data in the model from its original form as a matrix of X, Y, and Z values to raster grid-cell structure, contour lines, and even vector topographic data models.

Representing surfaces in a vector model

Trying to represent surfaces in vector models is a bit more complex than in raster models, but the basic idea for this model has been around since the early 1960s. Based on the idea that many parts of a surface are relatively flat, the programmers constructed a model that looks much like how three-dimensional computer characters (avatars) are constructed.

The primary vector model that represents surfaces in the computer is called the Triangulated Irregular Network (TIN), shown in Figure 5-11. The words in the TIN's name really explain what it does:

- **Triangulated:** The locations (X, Y, and Z coordinates) are created through a process of triangulation (a method of surveying based on the trigonometric properties of a triangle) and are represented as a set of non-overlapping triangles.
- **Irregular:** These triangles are irregular in shape because each represents a flat portion (facet) of the overall surface.
- **Network:** The triangles are joined together edge-to-edge to create a surface.

Each triangle is like the facet of a crystal. In this case, however, the crystal is a portion of the topographic surface. The three *vertices* (points of the triangle) of each *facet* (non-overlapping triangle) each have X and Y coordinates, and each has a unique elevation value. The slope (angle from horizontal) and aspect (direction) are uniform and constant for each individual triangle.

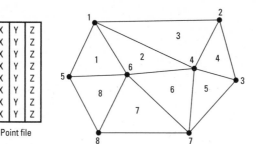

Figure 5-11: A Triangulated Irregular Network model.

TINs are topological data models, just like DIME and TIGER models presented earlier in this chapter, but they're specifically designed to incorporate the Z value component, whether it's topography or economic data. Like in other topological data models, the TIN data model contains explicit information about which features are related spatially to others.

The TIN model, shown in Figure 5-11, is pretty close to the model you see in Figure 5-7. Each model has a table of point values, and each has a set of polygons and links. In the case of the TIN model, the polygons are always triangles, and each triangle's vertices define it. The Triangle File table in the TIN model is much like the polygon table in Figure 5-7. Specifically, the Triangle File table shows

✔ The vertices that make up each triangular facet (the first three columns)

✔ The neighboring polygons (the second three columns)

The Point File table is different in the TIN topological model, as compared to the chain table in Figures 5-8 and 5-9, because it also contains the elevation (Z) value for each.

The TIN model has some definite positive features:

✔ It's incorporated in many vector GIS packages, so it can be used immediately for surface representation.

✔ You can easily convert it to matrix models, such as the digital elevation model (DEM).

✔ It allows for interpolation of missing Z values by using a wide variety of algorithms.

✔ You can make cross-sectional profiles of terrain and perform many advanced functions related to topography.

Chapter 6

Keeping Track of Data Descriptions

*N*o matter what type of geographic setting or scales of geographic data you use in GIS, you need to link the graphics with meaningful descriptive data. When you perform geographic analysis, you search, manipulate, compare, and analyze this data. Every GIS software package must manage descriptive data and link them to their respective objects on the map. Without this capability, your GIS is nothing more than a digital picture, and you may as well work with a paper map.

You can find many programs that contain both graphic elements and descriptions, such as graphics drawing packages, complex drafting and design programs, facilities management packages, and computer-assisted mapping programs. These programs aren't, technically, geographic information systems, but they may possess some of the functionality of GIS and definitely link data descriptions with graphics in a similar way. Getting a feel for how these simpler systems handle data-to-graphics linking can help you understand how more complex GIS do the same thing. Consequently, this understanding can help you select the software that meets your needs.

In this chapter, I describe how simple software systems store descriptive data and link them with the appropriate graphics. I tell you what these various systems do well (for example, computer-assisted cartography programs do a great job of storing geographic information and outputting maps) so that you can have an idea of how well the software can handle the analytical tasks that you have in mind. Who knows? You may decide that one of these simple systems fits your needs just as well as a full-blown GIS.

Knowing the Simple Systems for Tracking Descriptions

There are many kinds of software for dealing with different forms of graphical data like house plans, satellite images, and maps. In fact, development of these innovative programs took place simultaneously and for different purposes, as follows:

- **For analyzing satellite images:** People needed ways to manipulate electronic images gathered remotely from satellites orbiting the Earth. They began developing *raster manipulation* (image processing) software to analyze those digital images.

- **For basic mapping of Earth features:** Other folks were interested in mapping Earth's features, so they developed raster and vector mapping software.

- **For drafting and design work:** And people involved in drafting complex buildings, machinery, and so on created computer-aided design (CAD) and drafting software.

You can still find each type of software in the preceding list available today in one form or another. Each software type has its specific application, and various developers are collaborating across applications to the mutual benefit of all the different types of users. Because of the convergence of the different tasks, the properties of CAD programs have changed from their original form to a more GIS-like software. The descriptions in the following sections are somewhat simplistic, reflecting the original purpose and design of each type of software. When you consider your options, pay attention to the original conceptual design, as well as the capabilities that the software possesses in its current state.

Understanding computer-assisted cartography

Like GIS, computer-assisted cartography (CAC) was created to store, edit, and output maps of the Earth and — crazy as this sounds — other planets. CAC doesn't have the analysis capability that's at the heart of GIS, however. CAC is a straightforward process that moves from input to storage, retrieval, editing, and finally, output (see Figure 6-1). A CAC system just needs to store the *attributes,* or descriptive data, and have labels attached to the attributes so that those descriptive labels can be applied during output.

Figure 6-1:
The
computer-
assisted
cartography
linear
process.

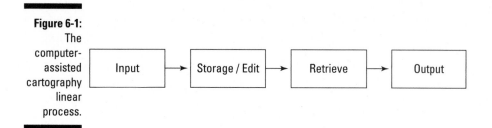

This simple system of linking the graphics to individual attributes makes CAC an efficient approach to store and output maps that have symbols and legends, but it limits both the amount of attribute data you can store for each entity and your ability to manipulate it. The usual way that such systems store attributes is as a list in which each *attribute* (description) is associated with individual graphic objects. For example, a graphic polygon might have the name Mirror Lake associated with it in the attribute list. Each map has its own geographic elements, called *entities,* and its own attributes. CAC doesn't allow maps to share either their entities or their attributes with other maps in the same database. As a result, you can't use CAC to overlay one map on another or combine their data in other ways.

CAC software allows you to translate input coordinates (such as digitizer inches) into scaled map data. You can also use CAC to convert from flat, planar coordinates to geographic coordinates, which allows you to convert map projections to land-based coordinate systems and vice versa. If your purpose is to input, archive, update, and output maps, CAC is the system for you.

If you plan to use a system only for map production, computer-assisted cartographic systems are both more efficient and more cost-effective than the more complex GIS software. CAC doesn't have the analysis capability that's at the heart of GIS, however.

Using computer-aided design

Like computer-assisted cartography, computer-aided design (CAD) isn't intended to manipulate or analyze cartographic data. In fact, CAD isn't intended to work with cartographic data at all — but it can. CAD systems — which are designed to store complex drawings in layers — store *attribute* (descriptive) data as linked lists tied explicitly to the drawing's *components* (the entities) on a layer-by-layer basis. This layer-by-layer division means that the data are not shared from one layer to another.

CAD is capable of performing visual overlay because of its inherent layered data model — unlike CAC, CAD is structured to store multiple data descriptions that apply to the same entity. Many times, the graphical overlay operations produce graphics that help you visualize the data, but they don't create new data from the various combinations of layers. GIS is designed to do so, however, which allows you to compare one map pattern to another.

Another major difference between CAC and CAD is that, generally, CAD doesn't include much capability to change projection or coordinate systems. You can't convert from one form of map projection to another or change from one coordinate system to another, for example.

One nice feature of CAD is that a standardized exportable version of the typical `.dwg` file (a common CAD drawing file), called the drawing exchange format (`.dxf`), is often easily translated from CAD to both CAC and GIS software. In fact, many people use CAD software for input because they find the user interface for the input task easier to use than the GIS interface.

Exploring raster systems

The basic raster system method of handling attribute data is to store them as a set of *ASCII characters* (meaning letters and numbers), usually numbers. Each grid cell has a single numeric code associated with it. In most raster systems, the numbers are whole numbers. The code, in turn, is linked to a legend description (the names of the categories) and to a symbol set (the colors or shades used to represent the categories) for final display.

Storing attribute data in a raster system

Storing attribute data in a raster (grid) system requires that you store the value of each individual square based on its location in the sequence of grid cells. So if you are using a grid system that starts in the upper left and moves to the right, the first value you store would be for upper-leftmost grid cell. You assign the numbers (codes) based on how you define the attributes from the start. While the values may be actual measurements for things like elevation or slope, you may also assign numbers to represent categories, such as 0 for water, and 6 for agriculture, and you make these decisions before you start putting the numbers in the database.

Suppose that you want to use the number 6 to represent agriculture. To do so, you place the value (code number) 6 in the first column, first row. It turns out the next cell in sequence is also agriculture, so you put the number 6 in the second column, first row. If you know that the third grid cell over (the third column) in the first row is water and you want to use the code of 0 for that, you just enter 0 in that grid cell (third column, first row).

You don't need to input these grid cell values manually anymore, but this example gives you an idea of how a typical GIS user would sequence them. You know the location of each grid cell of each type because you placed them where they needed to be, and you can easily retrieve the categories that you want based on the numbers you assigned.

When your GIS represents topography (the depiction of land surfaces), it doesn't have to use whole numbers. Some systems use only whole numbers to represent the Z value, but that approach limits the accuracy of the surface data. Most modern GIS software allows for the more accurate use of rational numbers expressed as decimals. The more accurate the numbers, the better are the results of your analyses. Check out Chapter 14 for more about working with topographic surfaces.

Map modeling with grid systems

Although the raster method is pretty simple to understand, you can't beat it when you want to store, analyze, and compare map attributes. Each grid cell for each grid layer is a uniform size and shape. Each grid layer is composed of identical numbers of columns and rows, in which each respective grid cell is positioned in exactly the same *geographic* (column and row) space (see Figure 6-2). In other words, a grid cell in the upper-left corner of one map layer is in exactly the same position as any other upper-left-corner grid cell for any other map layer.

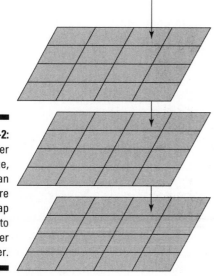

Figure 6-2:
In raster storage, you can compare each map layer to every other map layer.

Because each grid cell contains attribute data, you can easily retrieve the data and compare it to its respective grid cell in another map layer. You use this property when you compare one map to another (especially when you use the Map Analysis Package, or MAP, data model, which I cover in Chapter 5). In Chapter 17, I show you how this characteristic was used to create a complete GIS modeling language called *map algebra*.

Because raster grid cells are uniform in size and identically placed from one map layer to another, you may find them especially useful when you must perform map overlay operations for map analysis. Also, because the raster model is so simple, it makes map overlay faster than in vector systems, which usually have more complicated data structures (see the section "Managing data in Vector GIS" for more about these structures).

Working with Tables and Database Management Systems

Database management systems (DBMS) were created to manage data. As glib (and completely obvious) as that may sound, the idea is significant because DBMS weren't originally designed to manage all types of data, and specifically, not spatial data. The original database management system was a separate technology that had a different clientele in mind. In the early development of GIS, the DBMS stored attribute data, and the graphic data were stored elsewhere.

Environmental Systems Research Institute (ESRI) first fully implemented the link between graphics and descriptions stored in a database management system in its premier product called ArcInfo. The name is rather telling. Arc is the portion of the software suite that deals with the graphics, and Info is an existing database management system that's linked to Arc so that changes in *entities* (the graphics) are reflected in changes in *attributes* (the data) and vice versa.

Structuring simple relational data

At the heart of these systems is the database management system itself. These systems are called *relational database management systems* (RDBMS) because of their ability to form associations *(relations)* among the various data they contain. The data you search are contained in rows called *tuples* (pronounced *tooples*) and are grouped with corresponding rows called *relations*. These groupings of rows are called relations because the information they contain is related. Specifically, each piece of data (column) in a single row is related to that column's data in the other rows because each row contains identical categories that can be compared to one another. Each column represents an attribute — such as a parcel ID or house number for an address book — that's common to each tuple in the entire relation.

As an example, Figure 6-3 shows a land-use table. Each row contains a different Parcel-ID so that all the information in that row belongs to that one person. Each column contains a unique type of information, and each individual item is linked to the name in its respective row. The relationship among the pieces of data in a single row is a *tuple*. Collectively, you call this a *table of relations*.

PARCEL-ID	LUCODE	LANDUSE	STATUS	OWNER	PRICE
F2300223	010002	row crop	active	Sylvia Porter	$1,000
F4139992	020001	orchard	dormant	Robert Orange	$1,500
F2889881	030001	rangeland	active	Joseph Barney	$650
F9887632	010002	row crop	active	Sylvia Porter	$1,200
F0998944	010004	garden farm	active	Alice Stalk	$2,000
F7966501	010002	row crop	dormant	Jordan Waits	$1,100

Figure 6-3:
A relational database management system (RDBMS) table.

Each table operates like a set (for example, a set of land-parcel owners, together with the parcels they own and additional information about those parcels). These tables follow certain rules to store and, more importantly, retrieve data from these sets, including the following critical rules:

- **A table can't have duplicate rows.** Imagine, for example, if a row represented your land holdings. You need to have the information for each parcel included only once. Each row in a table should represent a unique entry.

- **A table must be searchable.** Each row of data is unique, so to find a unique entry, you can use one or more columns (containing a specific type of data) to search the rows. You could, for example, search by owner's last name, an ID number, or a land-use type. You can much more easily use a single column to search the rows of data if you set up the table so that you know which column you want to use. This preferred column — often the first column and the one you most often use to do your search — is called the *primary key*.

To successfully find the right data, though, you need to make sure that the primary key has a value that you can search. Missing values in the primary key could result in duplicate rows of data (that's a problem because the computer program has no way to determine which row it should be looking at). And even worse, if the primary key has no value, there is no way to connect to the remaining data in the row. It would be like having an address book that was missing names.

A good rule for database management systems is to make your tables very simple so that you know exactly what you're going to search, what your primary key is, and what you expect to find. Focus on one or two queries that you want to make for each table. Ask yourself, "What questions do I want to be able to ask about the data in this table?"

Getting more complex with relational joins

Simple databases, although easier to understand and control than more complex ones, aren't going to be of any use to you if they don't give you access to the large, complex datasets like the ones that you encounter in most GIS operations. But a relational join can help by allowing you to segment your database into nice bite-sized chunks that you can compare later (as shown in Figure 6-4). The *relational join* is a linking mechanism that allows you to match data from one table to another. You can link as many tables as you like by matching a column (the primary key) in one table to another column in the second table. The linked column in the second table is called a *foreign key*.

LUCODE	LANDUSE	STATUS	OWNER	PRICE
10002	row crop	active	Sylvia Porter	$1,000
20001	orchard	dormant	Robert Orange	$1,500
30001	rangeland	active	Joseph Barney	$650
10002	row crop	active	Dabney Porter	$1,200
10004	garden farm	active	Alice Stalk	$2,000
10002	row crop	dormant	Jordan Waits	$1,100

COMMON KEY			
OWNER	STREET	CITY	Phone
Sylvia Porter	401 Main St.	Aggieland	555-5542
Robert Orange	1410 Oak Ave.	Aggieland	555-0024
Joseph Barney	12 Eagle Ct.	Pottsville	555-4429
Dabney Porter	401 Main St.	Aggieland	555-5542
Alice Stalk	222 W. Elm	Aggieland	555-2218
Jordan Waits	1002 E.Pine	Aggieland	555-0991

Figure 6-4:
A relational join links tables by a common column.

When you establish a relational join, all the information in the first table is shared with all the information in the second table, which is much easier than trying to make a huge, complex, comprehensive table from scratch. It's much easier to design small, individual, focused tables (that can be related to other small ones) than to create large, unwieldy ones. Using smaller tables also gives you flexibility if you need to add new types of data to your database. You simply add a new small table that contains a column common to other existing tables.

Relating tables with joins sounds easy, doesn't it? Well, it is, and it also adds flexibility and robust search capabilities to your GIS database. But you do need to be sure that the tables have at least one column in common. The common column enables the join and allows the tables to share information.

Managing data in Vector GIS

ESRI ArcInfo is a classic model of how Vector GIS employed the power of the database management system to store and manage all attribute information. Figure 6-5 is a simple graphical description of how the system looks. One set of programs (the Arc side) contains the graphics portion of the database (the *entities*) and the other side (the Info side) contains the descriptions (the *attributes*).

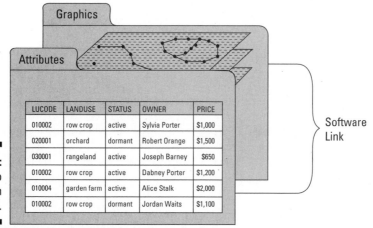

Figure 6-5: The ArcInfo system model.

LUCODE	LANDUSE	STATUS	OWNER	PRICE
010002	row crop	active	Sylvia Porter	$1,000
020001	orchard	dormant	Robert Orange	$1,500
030001	rangeland	active	Joseph Barney	$650
010002	row crop	active	Dabney Porter	$1,200
010004	garden farm	active	Alice Stalk	$2,000
010002	row crop	dormant	Jordan Waits	$1,100

Software Link

Database management systems may look like the tables of a spreadsheet, but they're actually much more complex, and they operate quite differently. For a system such as ArcInfo (or other Vector GIS that uses database management systems), the points, lines, and polygons all have descriptive information about them in the columns of the database management system, but they also have identification codes that link this information to the entities on the Arc side.

Figure 6-6 shows an example of a table from a recent version of ESRI ArcInfo, called ArcGIS version 9.X. This table includes polygon ID codes, as well as a lot of columns of spatial attribute data that you can use to search. The polygon codes allow the software to find all the graphics (in this case, polylines) that are associated with them. And each code is also linked to all the other elements in that row so the computer can find them as well.

FID	Shape *	L_F_ADD	L_T_ADD	R_F_ADD	R_T_ADD	PREFIX	PRE_TYPE	NAME	TYPE
1968	Polyline	1052	1078	1087	1113			Atlantic	Dr
1969	Polyline	1080	1142	1115	1143			Atlantic	Dr
1970	Polyline	1144	1198	1145	1183			Atlantic	Dr
1971	Polyline	1200	1218	1185	1225			Atlantic	Dr
1972	Polyline	1220	1238	1227	1251			Atlantic	Dr
1973	Polyline	1240	1278	1253	1279			Atlantic	Dr
1974	Polyline	852	864	889	927			Image	Ave
1975	Polyline	928	998	929	999			Image	Ave
1976	Polyline	860	898	861	899			Fielder	Ave
1977	Polyline	902	920	901	921			Fielder	Ave
1978	Polyline	922	946	923	947			Fielder	Ave
1979	Polyline	948	970	949	971			Fielder	Ave
1980	Polyline	972	998	973	999			Fielder	Ave
1981	Polyline	814	846	815	847			Curran	St
1982	Polyline	882	914	871	915			Curran	St
1983	Polyline	916	946	917	947			Curran	St
1984	Polyline	948	974	949	975			Curran	St
1985	Polyline	976	978	977	979			Curran	St

Figure 6-6: The database table for a spatial database in ArcGIS 9.X.

Storing data in Raster GIS

Raster GIS normally stores attribute data in the grid cells themselves. Each grid cell gets its own value for each layer, and you can easily locate and retrieve each cell using that value. But, like with Vector GIS, the need to hold more complex datasets and more descriptive information for each layer ultimately led to the incorporation of RDBMS inside Raster GIS. In a way, the data storage is not that much different than you find with Vector GIS, except that in raster, each grid cell category (rather than, for example, each polygon) has a set of descriptive information.

You query data in a Raster GIS with RDBMS in the same way that you do with a vector system. The tables look pretty much the same, and even the query tool is often identical, so you don't have to figure out a new one. This capability in a Raster GIS makes the computations simple and extendable to remotely sensed raster data. It also adds the RDBMS's power to store and manage large datasets.

Searching with SQL in any GIS

Modern GIS software relies heavily on the idea of storing descriptive information in a set of tables, but not just any tables — database management system tables. The database management system contains a powerful set of query tools because, like any RDBMS, it employs the Structured Query Language (SQL). The database contains data organized in tables *(relations),* and the SQL searches these data.

SQL is a fairly standard computer language designed specifically for accessing, querying, and manipulating databases. It allows you to search for very specific information, retrieve it, delete or add to it, update it, and combine queries. SQL searches database tables by using commands such as `Select`, `And`, `Or`, `Insert`, `Delete`, `Order By`, and many more.

The following steps show how to retrieve a subset of data by using the SQL Select command:

1. **Determine the subset of information that you want to retrieve from your GIS data.**

   Suppose you have a table of land attributes like the following (Land_Ownership), and you need only the information related to the land use and ownership categories.

Polygon ID	Land-Use Type	Land Condition	Public / Private
7255	Recreation	Active	Public
7721	Agriculture	Idle	Private
7246	Agriculture	Active	Private

2. **Structure a SQL SELECT command to specify the attributes that define the information you want to retrieve.**

   Because you are interested in only the land use and ownership categories, you specify those attributes and issue the following command

   ```
   SELECT Land_Use_Type, Public_Private FROM Land_Ownership
   ```

   where `Land_Use Type` and `Public/Private` are the attribute names and `Land Ownership` is the table name.

   That command returns the following subset of the original table:

Land-Use Type	Public / Private
Recreation	Public
Agriculture	Private
Agriculture	Private

This example returns two columns from the database and leaves the rest behind. This simple operation demonstrates that you can execute literally hundreds and thousands of other, more complex queries.

Understanding Object-Oriented Systems

An *object-oriented database management system* (OODBMS) extends the traditional database management system so that it can model and create objects based on those data. *Objects* are collections of data and computer code that have two primary characteristics — *inheritance* (a common set of properties that move with them) and *hierarchy* (a place within larger classes of objects that have common properties).

By using software that has this capability, you don't have to re-create the properties of some groups of data that act as objects; the object inherits those properties, which saves a great deal of time. This inheritance also prevents you from changing properties of data arbitrarily and making mistakes by combining incompatible data — like putting 2-inch pipe and 8-inch pipe together for a sewage line.

Storing attributes with object-oriented systems

If you inspect a *geodatabase* — the ESRI application of object-oriented database management systems for GIS — it looks, at first blush, pretty much like any other database. It has columns and rows, tuples and relations, primary keys, and (if necessary when primary keys are not common to two tables) foreign keys. In short, the user interface looks pretty much the same as any RDBMS-based GIS to the user. But the geodatabase's ability to store collections of data as objects makes it different and more powerful than systems that lack the inheritance property of objects in object-oriented systems.

The OODBMS has the same descriptive items that you normally find in a GIS implementation of the RDBMS. What gives this system its power and makes it a major innovation in GIS isn't in the way it stores traditional descriptive data, but in that it also has a set of tables that go beyond describing the qualities of geographic features — what they are, how big they are, and so on — by also describing how those qualities translate into real geography.

Using object orientation to enhance descriptive information

Every geographic feature, whether it's manmade or natural, has properties that affect its functionality. Superhighways have the ability to move more cars than two-lane roads, so superhighways are the better choice for constructing an interstate highway system. And some types of vegetation grow faster than others, so the fast-growers are often chosen for shelterbelts and ornamental use. These and the millions of other functional properties make the simple descriptions of content, color, size, shape, and so on less useful for decision-making. Through experience, you know how a feature's properties affect its use and implementation.

Now, imagine you have that practical experience stored explicitly in your GIS tables. Experience is translated into a set of rules and behaviors that exist for each object (for example, each road type). The biggest advantage of object-oriented systems such as the ESRI geodatabase is that they add these behaviors to the stored objects. So, if you try to put geographic features into a database that violate the known behaviors of the object class they belong to, the software rejects them. Also, the behaviors often translate into how you can and can't link different objects in your GIS.

Knowing the packaging descriptions for different objects

An enormous number of industries, organizations, and settings use GIS, which makes it an enormously powerful tool. But it also has the potential to be a rather large, even ungainly, piece of software because it tries to meet too many different user needs. The geodatabase steps in and allows for the creation of large, comprehensive sets of rules and conditions that largely depend on the types of data, conditions of use, and methods of geographic-feature interaction specific to each set.

In RDBMS terminology, a general data model that's modified to address the specific data and query needs for a given application is called a *schema*. So, for example, you might have a general schema for vegetation analysis that you can modify for desert vegetation. ESRI and its software users have created an ever-increasing array of industry-specific schemas that they refer to as *geodatabase models*. ESRI intended not only to share these schemas, but also to use them as examples for other industries when developing their own schemas.

Because ESRI wanted to share the geodatabase, the institute has developed best practices that act as a guideline for developing these geodatabase models, regardless of content. These best practices indicate that each model should include the following:

✔ A case study that implements the model as an example for new users.

✔ An import template into which users can import their own data.

✔ A white paper that explains the model.

✔ A data-model poster that provides a graphic idea of how all the data fit together.

✔ A set of tips and tricks for new users based on the case study.

✔ (Optional) Web-based collaboration with other users.

Table 6-1 lists well-established, industry-specific geodatabase data models created by ESRI and their customers. The table also indicates which of these data models, at the time of this writing, have case studies, templates, and a variety of tools for incorporation into the models.

Table 6-1	Industry-Specific Geospatial Data		
	Case Studies	*Templates*	*Tools*
Address	X	X	X
Agriculture	X		
Archiving	X		
Atmospheric	X	X	X
Basemap	X	X	X
Biodiversity	X	X	
Building Interior Space			
Census-Administrative Boundaries	X	X	
Defense-Intel	X	X	

	Case Studies	Templates	Tools
Energy Utilities		X	
Energy Utilities — MultiSpeak		X	
Environmental Regulated Facilities		X	
Fire Service		X	
Forestry		X	
Geology	X	X	X
GIS for the Nation	X	X	
Groundwater	X	X	
Health	X	X	
Historical Preservation and Archaeology		X	
Homeland Security	X	X	
Hydro	X	X	X
International Hydrographic Organization (IHO) S-57 for ENC	X	X	
Land Parcels	X	X	
Local Government		X	
Marine	X	X	X
National Cadastre		X	
Petroleum	X	X	
Pipeline	X	X	
Raster	X	X	
Telecommunications		X	
Transportation	X	X	
Water Utilities	X	X	X

These geodatabase data models can get quite complex because of the number of geographic features that users must consider, the properties for each of those features, and the potential number of rules for interaction. The posters that accompany many of these existing models provide an excellent graphical view of their operation and organization. Figure 6-7 shows a small portion of a poster for the water utility model from Table 6-1.

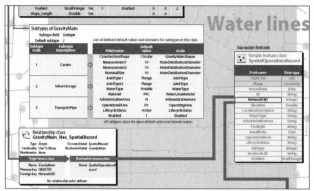

The geodatabase model and other object-oriented spatial data models are likely to grow in number and complexity. They add tremendous robustness and accuracy to an application, but using them means that you need to know more about your particular industry. In many applications, you can find far more missing relationship rules than rules that are completely understood. Advances in what people know about the geography that drives an industry can result in more complete sets of rules, as well as far better models built on these OO databases.

Chapter 7

Managing Multiple Maps

*1*f a picture is worth a thousand words, a map is worth a million. With GIS, you can work with more than one map at the same time. So, GIS gives you many millions of words at your fingertips. Every portion of the Earth, whether it's natural or man-made, is very complex. Even if you ignore for a minute that every portion of land has depth, you can map many possible features, including houses, people, insects, plants, diseases, pollen, sidewalks, gas lines, salaries, streets, temperature, and pressure. The list is endless.

When mapped, each of these features provides a portion of the geography of a piece of land, and each set of similar features represents a separate theme. Within each theme, the explicit locations of each object, combined with the thousands of possible implicit spatial relationships, tells volumes. But, like any story, your land needs many characters — many themes. To get a complete picture, you need to have many mapped layers.

ESRI provides software, lessons, and technical support for K–12 Education around the world as a means of encouraging spatial thinking. The K–12 educators at ESRI call the map-layering process "making an Earth sandwich." And, like an Earth sandwich, a GIS works only when it has more than one layer. The individual layers become useful only when you can extract information from each and compare and contrast that different information. To make this comparison, the software must be capable of storing, retrieving, editing, analyzing, and producing new data from all layers. In this chapter, I show you how the various types of GIS systems do all these data management tasks.

Layering Data in GIS Models

Geographers have long recognized the many types of geographic features and that many of these features tend to occur in similar patterns. So, for example, if you look at where most of the people in the world are located, you also see a strongly similar pattern of water sources, such as coastlines, lakes, springs, and rivers.

These similar patterns may or may not be related, may or may not show cause and effect, and may or may not occur at the same time. But they all definitely do one thing — they demonstrate that the more of these patterns you map, the better you can understand the geography.

Each GIS layer must exist as a separate searchable item that you can access independent of the other layers. And, at the same time, the GIS must also have these characteristics:

- **Data segmentation:** After you retrieve layers, the individual geographic features and the related data that make up each layer must also be readily accessible.

- **Data comparability:** When you have access to the individual layers, features, and their descriptive content (categories and values), you must be able to compare these categories and values across any and all layers, and any and all content.

Accessibility and comparable data relationships are the heart of GIS — and they're what give GIS much of its power.

In Chapter 16, I show you a set of powerful techniques that allow you to compare layers to one another. These techniques, collectively called *overlay,* come in many forms and require you to compare the data by using various methods, such as formal logic and mathematical manipulation. You use these techniques to create completely new layers that you can then analyze.

Comparing the Map-Handling Capabilities of GIS System Models

There are three basic types of system models for GIS: hybrid, integrated, and object-oriented. Each model type is designed to manage multiple layers. But each takes a fundamentally different approach to storing and relating geographic data. You don't often see the internal workings of the GIS, and I don't go into unnecessary detail. You may want to choose a GIS software package for your organization sometime, so having a basic idea of how overlays work will help you in your selection process. Because how you work with overlay operations affects which GIS model you use, you need an understanding of the similarities and differences, strengths and weaknesses of each model to help you make the right choice.

Checking out a hybrid system model

One successful method of managing many different layers inside GIS is through a hybrid systems model conceived in the early 1960s during the development of the first operational GIS. The *hybrid* model structure recognizes the fundamental difference between graphics features *(entities)* and their descriptions *(attributes)*. Many of the original vector-based GIS packages used the hybrid model, including the original ESRI ArcInfo software. See the sidebar "Combine and create with ArcInfo," in this chapter, for a bit more about features and attributes in this hybrid system.

What makes this model a hybrid is that the two parts — the graphics (entity) part and the description (attribute) part — exist entirely separately from each other (see Figure 7-1). When you search a hybrid system, you retrieve the graphics part, meaning the geographic features, separately from the data part (the attributes for the features), but they are connected through software that communicates between these two parts.

In the original ArcInfo software, a retrieved layer was called a *coverage* because its primary function was to overlay other layers.

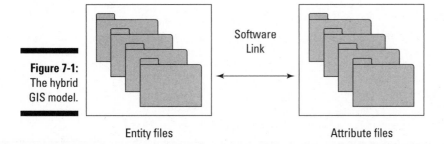

Figure 7-1: The hybrid GIS model.

Software Link

Entity files   Attribute files

The entity files that make up the retrieved layer contain identification codes; feature types (point, line, or polygon); and (typically) information about the number of features, their measurements, and sizes, all captured during the input process. The entity files connect with a set of computer code — called *links* or *pointers* — that literally point back and forth between each geographic object in the entity files and its description in the attribute files. These attribute files are stored within the tables of a traditional relational database management systems (RDBMS), which I discuss in Chapter 6.

When you retrieve and overlay two layers in a hybrid system, you get a third layer that combines the features and data of the other two. This combination requires the software to "rebuild" the entity and attribute files to match the changed topology of the third layer. Systems that work this way are called *topological graphics data models.*

The rebuilding process that these models use compares the new entity records in the one set of files with the attribute records in the other and then reworks the data to check, correct, and rebuild topological relationships. The system also needs to perform this topological rebuilding when it reconfigures the graphics of a single layer.

Different approaches to these systems, called *integrated topological models,* perform the rebuilding operations more or less simultaneously with the process of overlay and other analytical operations affecting topology, which makes your job somewhat easier because you don't have to wait for the topology to be rebuilt.

Combine and create with ArcInfo

In some ways, the ArcInfo model is similar to the Map Analysis Package (MAP) model covered in Chapter 5 because each layer has a name (for example, Vegmap or Soilmap). The name provides access to each layer individually. When ArcInfo was popular, the command-line user interface addressed combinations of layers and their attributes through a natural language structure, again much like the MAP model. So, you could overlay layer1 onto layer2 to get layer3.

ArcInfo, a classic example of the hybrid system model, was created as a *topological graphics*

data model. When overlay operations combine the graphics of one layer (layer1) to those of another (layer2), the polygons themselves become unique combinations of the two originals. By combining layer1 and layer2, you create a totally new layer3 that must also have its appropriate entities, attributes, and topological structure. Layer3 now has new attributes that reflect the combination of layer1 and layer2 attributes, and those attributes are assigned to the new layer's features (the unique polygons).

Eliminating pointers with integrated system models

Unlike hybrid systems, integrated systems treat entities, topology, and attributes all the same by storing them in the same sets of files. Integrated systems still treat each layer as a separate set of data and provide a naming mechanism so that they can find any layer needed. But the integrated systems don't need links or pointers from entities to attributes because all the records are contained in the same RDBMS files (see Figure 7-2). Like the RDBMS itself, the integrated system simply connects one set of tables to another through relational joins.

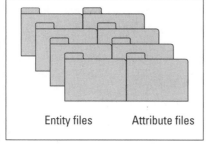

Figure 7-2:
The integrated GIS system model.

Entity files   Attribute files

Most integrated systems were originally derived from other forms of spatial software that were more closely allied to facilities management and computer-aided manufacturing. Their original intended use was for visual overlay, not managing multiple maps for analytical combination. If you were doing facilities management, for example, you'd want to know the location of specific pipes, fittings, access points, joints, and electrical wires and conduits relative to one another. A visual overlay could show you this positioning.

Although integrated GIS systems weren't originally designed for overlay operations, they've evolved to perform such tasks quite effectively. One major difference between integrated and hybrid systems is that integrated systems don't need pointers between two sets of files. If you plan to use your GIS to manage many layers that have limited overlay operations, the integrated system is a good choice.

Getting better control with object-oriented system models

The ESRI geodatabase model (covered in Chapter 6) is built on the hybrid concept (see "Checking out a hybrid system model," earlier in this chapter) but possesses a unique and powerful object-oriented nature that makes handling layers, whether individually or in multiples, a bit more complex during database building. *Objects,* the groups of geographic features represented in the database, have a unique property called *inheritance,* which means

✔ **Objects can exhibit properties and rules that govern their behavior in your database.** These properties and rules make the features you include in your GIS more realistic in their representation of geography.

✔ **Any feature that belongs to an object class (a *class* is a group of similar objects, such as streets or land-use polygons) shares (inherits) the properties of the class.**

✔ **An object class can also belong to larger classes of objects that also share their properties.** So, a group of forest polygons would share the properties of all forest polygons, but they would also share the properties of a larger class of polygons — vegetation polygons (see Figure 7-3).

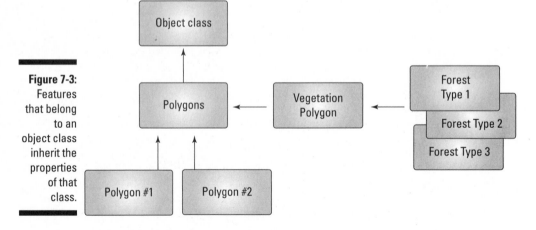

Figure 7-3: Features that belong to an object class inherit the properties of that class.

Class properties can include unique properties, such as your naming conventions, and even some unique rules of behavior. Class properties also prevent improper combinations through overlay and non-overlay operations alike. Rules prevent you from doing things such as creating GIS plans for building roads over rivers where no bridges exist. They protect you from creating mathematical inconsistencies by improperly combining layers of different data measurement level, for example, by multiplying numbers that represent nominal categories by those representing ordinal, interval, or ratio scales of data measurement. (See Chapter 1 for an explanation of these data measurement levels.)

Having these built-in protections presumes that the governing rules exist somewhere in the database. Most object-oriented systems allow for a separate set of tables that contain these specific rules and logic. In that sense, such systems resemble the hybrid system more than the integrated one (I describe both system types in the preceding two sections). The good news for you is that the user interface doesn't really depend on the system model so you can move from using one model to using another model seamlessly.

If you want to have some serious control over how your map variables interact (and if you know more than a bit about how the geography controls these interactions), an object-oriented system is your best choice because you can explicitly include your knowledge as rules and properties in your database.

Opting for an Object-Oriented Model

The speed and efficiency of modern computers have fundamentally altered the way computer systems, even large ones such as GIS, work. User interfaces have moved from command-line entry to graphical user interfaces; you can store large amounts of data on massive storage devices; and (most importantly) you find little difference in how programs that have different speeds and performance work on your system.

So, speed and efficiency don't really play a part in your decision making, and choosing a system model has much more to do with what you want to accomplish and the nature of your analysis. These three major factors might help you decide on an object-oriented system model:

- ✔ **How well you know your data:** If your system is based on a well-known geography (for example, on your knowledge of how crime occurs geographically), the object-oriented model allows you to code that logic into your database.

- ✔ **What depth of analysis you require:** If you need to supply additional geographic rules and logics for some existing geographic features that you understand well, then you should use an object-oriented system model that allows you to include those rules.

- ✔ **How you see your GIS needs evolving:** Moving toward the object-oriented GIS can help your system become more adaptable to innovations in system design that will continue to use object-oriented programming and database management techniques.

If your industry can use object-oriented data models (which I talk about in Chapter 6), take advantage of those models when you can. The vast majority of object-oriented data models are geodatabase models that work with ESRI ArcGIS software. Consult with others in your industry before you decide on a system.

Although object-oriented systems are powerful and provide a much better geography than simpler systems, you don't have to use that power if you don't have an immediate need for it. Many users of object-oriented GIS software never use the full potential of the object-oriented model, just as most folks never use the full potential of their word-processing software. In any case, the model's power is there when and if you do need it.

Chapter 8

Gathering and Digitizing Geographic Data

. . . .

In This Chapter

▶ Using the best data for your GIS

▶ Incorporating data from GPS, remote sensing, field collection, and censuses

▶ Adding existing maps to your GIS

. . . .

*N*o GIS can operate without substantial amounts of quality geographic data, no matter how you intend to use it. You can purchase these data from commercial vendors, but the surest way to guarantee your GIS database integrity is to build it yourself or carefully supervise its construction for the specific task or tasks that you have in mind. You know exactly what you want your GIS to do and how you want it to help with your decision-making.

Because GIS offers so many possible applications, many possible types of data exist that you can put into the GIS. This chapter discusses a few examples in four basic data groups that you're most likely to encounter — GPS, remote sensing, field data, and census data. In this chapter, I also show you how to work with existing maps and prepare them to be digitized.

Identifying Quality Data

The sources of GIS data, as well as their accuracies, sampling methods, and many other factors, affect not just the quality of a GIS database, but also how effectively it serves your decision-making needs. I've personally tried to build GIS databases based on over 10,000 observations of wild animals. Unfortunately, because the study was based on behavior, rather than movement through the study area, the researchers took the vast majority of the samples in front of several den locations, making those samples unacceptable for GIS implementation. Because the researchers collected all the data at one spot, while the data (animals, in this case) wandered all over the study area, the samples don't reflect the spatial nature of the data. I've had similar problems because of data that has incorrect classifications, scale problems, timeliness issues, and so on.

If you want GIS to be useful for finding solutions to geographic problems, it needs to be appropriate (to the degree possible) in the following ways:

- ✔ **Scale or resolution:** How much detail do you need for your study?

- ✔ **Measurement level:** Do you need interval data, or are categories enough?

- ✔ **Accuracy:** How well can your measurement tool capture your data?

- ✔ **Sampling method:** Do you collect all the data in all the places you need?

- ✔ **Timeliness:** Do you work with time-sensitive data that change quickly and need to be collected right away?

- ✔ **Data type:** Are the data the appropriate data for you application, both in subject matter and in format? (Do you need field data or satellite data, for example, or do you need soil data rather than temperature data?)

- ✔ **Data classification system:** Do you use the same data classes as other layers in your database (for example, land-use classes from 1955 versus 2005)?

- ✔ **Completeness:** Have you collected all the data that you need to answer your question?

Select only the best-quality and most appropriate data type for your needs. Collecting more data than you need and gathering data that includes irrelevant details add to the costs and reduce the effectiveness of the resulting GIS.

Importing Statistical and Sensory Data

To keep things simple, I've divided GIS data into two types: statistical and sensory. This division is a bit artificial. I'm defining *sensory data* as those data most commonly associated with distant sensing devices, such as the Global Positioning System (GPS), and various forms of imagery, including both aerial photography and digital satellite data. *Statistical data* include field data and census data, both of which usually rely on some form of direct contact by a person to collect.

Getting information from GPS data

Folks often mistake GPS for GIS, partly because the two acronyms sound so much alike and partly because GPS is more common as a consumer product. Using the advanced technology of GPS to provide highly accurate location data is a huge bonus when building a GIS.

GPS requires a large number of satellites orbiting the Earth (32, as of this writing), strategically placed base stations, and receiving equipment (shown in Figure 8-1). By triangulating radio signals sent from these orbiting satellites down to Earth, geographers and geodetic scientists can calculate the latitude (north-south location) and longitude (east-west location), as well as elevation data, for virtually any location within the satellites' lines-of-sight.

Figure 8-1:
A typical
GPS
receiving
station.

Because GPS requires line-of-sight, it isn't as useful for places such as the tropical rainforest because the trees prevent the receiver and the satellite from seeing each other. Even in conditions without the problems of tree canopy, for best results, it takes three or more satellites and some expensive receiving equipment (what the stores call GPS units). But even inexpensive receivers can yield some incredibly accurate results (see Figure 8-2).

GPS receivers that cost less than $100 can still give you accuracies of less than 10 meters, depending on the number of satellites your unit can see at one time.

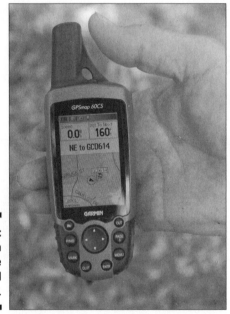

Figure 8-2:
An
inexpensive
handheld
GPSreceiver.

Using remote sensing to create maps

Scientists have used aerial photographs to map portions of the Earth for nearly 100 years, and although the technology today is mostly digital rather than film, it's still available. Most of the U.S. Geological Survey's original topographic maps were derived directly from observations from aerial photographs. In fact, USGS created or improved a large majority of map products derived before the mid-1970s by interpreting these pictures.

When satellites sense the Earth, they detect radiation values for a large grid of small squares covering part of the earth's surface, much like the grid cells described in Chapter 4. Because these grid cells are specifically related to recorded imagery, they're called *picture elements* — or *pixels,* for short. Each time a sensor senses an area, it records a single radiation value in a particular radiation band for each pixel. The more sensors available on a single satellite, the more sensor bands you pick up because each sensor looks for a unique part of the electromagnetic spectrum (see Figure 8-3). So, rather than a single set of digital values representing only one radiation band, you get multiple sets of values, each based on its own part of the spectrum.

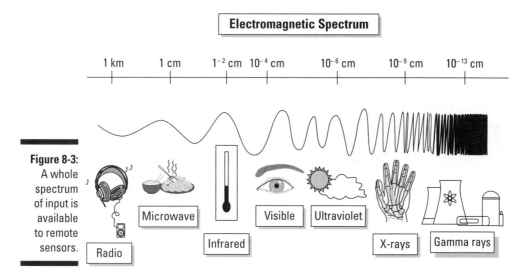

Electromagnetic Spectrum

1 km | 1 cm | 1^{-2} cm | 10^{-4} cm | 10^{-6} cm | 10^{-9} cm | 10^{-13} cm

Radio | Microwave | Infrared | Visible | Ultraviolet | X-rays | Gamma rays

Figure 8-3: A whole spectrum of input is available to remote sensors.

Enhancing images

With layered digital images, you can work with the separate spectral (wavelengths of radiation) layers and decide how to put them together to get the exact picture you want. For example, when you take photographs with a digital camera, you can change the appearance of those pictures by using your computer (or sometimes even the camera's software). You can change Grandpa's red eyes to their original brown or get rid of that line on the image that resulted from a hair that fell across the lens. You adjust the picture in this way, called *image enhancement,* to make the image clearer or simply more understandable to image analysts.

You can use images gathered from aerial photographs as *base maps* (a starting point for more complex maps) when you're heads-up digitizing (digitizing off the screen) because these photographs give a nice, relatively current picture of the Earth. Also, enhanced images make heads-up digitizing even easier than the raw images do because enhanced images allow the human eye to see things not visible under normal circumstances. For example, you can make lines (such as fences) and boundaries (between fields and pastures) stand out so they can be easily seen for digitizing. This sounds a bit like cheating, but it really isn't. It's a bit like what studio photographers do to create those special-effects pictures for weddings, graduations, and other special events.

Classifying images

Scientists classify satellite images, often with the notion of putting those images into a GIS package. Rather than relying on the human eye to do the work, the computer classifies the images in two basic ways:

- **Supervised classification:** With supervised classification, you first identify an object you know, such as a lake. You then tell the computer to look at the pixels that you selected and find other pixels that have the same kind of spectral values (brightness values in each spectral range). That's the non-technical explanation. The actual computer program is all about statistics, but it really just tells the computer to "find other stuff that looks like this selection and call that stuff a lake."

- **Unsupervised classification:** If you want to be a bit more rigorous or you're unfamiliar with an area, you can use unsupervised classification, in which the image processing software randomly picks different pixels as starting points. You can include a class estimate (such as the upper and lower limits) of the types of features you see in an image. The software groups the pixels into those classes and then searches to find others like them.

    For example, if you want the computer to pick ten classes, you get a random set of pixels grouped into ten classes. Then, the computer finds pixels that "look" like (have the same range of pixel values) each group and classifies them. After you review the results, you can tweak the upper and lower limits of your class values as needed to capture all the class information you want.

    This approach may not pick up all your pixels the first time, or it may get classes confused. In general, if you put it through a few trials, it can find a pretty good set of classes. But it doesn't give those classes names — that's your job. Nice to know you're needed, huh?

Whichever classification method you choose (supervised or unsupervised), you get what remote sensing scientists call an *image map*. And you can put that image map into your GIS. In fact, one of the advantages of the raster GIS is that its data structure is nearly identical to the data structure of these satellite images. Figure 8-4 shows a typical image map.

Figure 8-4:
A typical digital satellite image is a set of pixels.

When you use remotely sensed images, you have access to a large number and different types of land-sensing satellites, so you can obtain new images frequently. As a way of updating GIS data, especially information such as land cover and land use, this type of GIS input is invaluable because it covers so much territory, so frequently, and at a relatively minor cost per unit area.

Collecting field data

People collected data in the field long before satellites, GPS, or even GIS were around. Assembling field data can involve conducting house-to-house surveys; collecting traffic data along roads; recording the air temperature and other atmospheric data; or gathering soils, vegetation, insects, or any number of other environmental samples. For hundreds of years, naturalists, anthropologists, and many others collected data about people, plants, animals, landforms, and the like, but collecting such data with the idea of subsequently producing maps came a bit later.

No matter what types of data you intend to collect, you generally can't find and record every single instance of the features you're examining. Imagine, for example, that you want to study grass in your backyard. You're going to be there a very long time if you intend to collect data on every blade of grass. Or perhaps you want to measure the temperature everywhere in your city. Because temperature is *continuous* (it occurs in virtually every possible place all the time), it's physically impossible to collect temperature everywhere. In each case, you're forced to collect data from a sample of the total. For GIS, you need to sample geographic space (which is, by definition, continuous).

Maps represent space, so the samples that you take (spatial sampling) must represent that space accurately. That's a fancy way of saying if you're sampling trees on an island, you can't sample on just one side of the island. You need to sample everywhere that might possibly feature trees to get meaningful results for mapping. Lack of spatial sampling is a common problem with many GIS databases.

When you take samples of data, be sure that you take samples from all sorts of different places — places that have different sizes, shapes, elevations, and climates.

By using the sampling process, you're trying to determine the numbers, distributions, and locations of geographic features in different parts of your study area. You can sample spatial data in several ways, depending on whether you're trying to collect point, line, or area data. These data can be grouped into the following classes, as shown in Figure 8-5:

  ✔ **Clustered:** Sampling focuses on areas that have a lot of features from which you can sample.

✔ **Systematic:** You use a specific, often regular, pattern to sample. For example, you sample at every meter along a line.

✔ **Random:** The sampling has no pattern at all. This approach often uses a computer to generate random numbers so the randomness of the samples has no human bias.

You can also *stratify* these samples (divide them into groups, or *strata*) so that you can do a specific portion of your sampling in each of a number of sub-portions of your sample area. To stratify your sample of who watches certain television programs in your city, you could divide the city into sub-portions, or neighborhoods. Then, you pick a certain number (for example, 25 people) in each neighborhood to sample randomly, systematically (for example, every fifth house), or clustered (such as where housing density is highest).

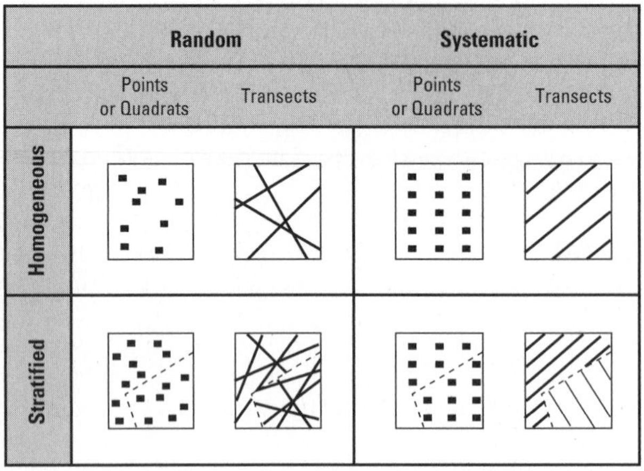

	Random		Systematic	
	Points or Quadrats	Transects	Points or Quadrats	Transects
Homogeneous				
Stratified				

Figure 8-5: Spatial sampling schemes for point, line, and area data.

Spatial sampling can give you a much better representation of the geography. Whatever method you use, the key is simply to recognize that your data has a spatial distribution and a sampling strategy that ignores that distribution usually results in less than useful data.

Working with census data

Almost by definition, a census is a complete counting, although that's seldom possible because (for a census of human population) parts of the population may have no address, may be purposely hiding, or may simply be missed. There are many types of census, including censuses of agriculture and bird populations, that don't involve counting humans.

The most common type of census is the census of population. The United States performs a census every ten years on years ending in zero. This census results in the most common census products: The census DIME files and the subsequent TIGER files (see Chapter 5 for more on these file types), both of which are readily incorporated into nearly all modern GIS software. These files, however, are just the entities. Over the years, not only has the Census Bureau developed a substantial set of attributes, but it has also created an entire census geography used by human geographers, as well as business people, government bodies of all levels, and those who want to include socio-economic data into their own GIS activities. Many other nations have censuses, and you can often find their data available for input into a GIS database, as well.

The attributes that are linked to DIME and TIGER files are found in *summary tape files* (STF), which have been around since the old days when the data were stored on 9-inch magnetic tape reels. Generally, the STF contain population demographics, including schooling, ethnicity, number of children, employment, income, housing, and even number of vehicles. These data are in map-ready form, from the 1980 census DIME files to the TIGER files used today. The DIME file structure is still used because the data collected for the 1980 census are still useful and are stored in that structure. (See Chapter 5 for a refresher of the differences between DIME and TIGER files.) Additionally, with some work, you can convert earlier census figures to GIS-compatible forms. Details about the content of the STF are readily available from the Census Bureau at

```
www.census.gov/geo/www/index.html
```

Getting Existing Map Data into the Computer

Thousands upon thousands of hard-copy maps have yet to make it into computers. You can digitize hard-copy maps in three basic ways: manual or hand digitizing; scanning; and heads-up digitizing from a scanned image, using your computer monitor. After you choose a method, you need to follow a process to prepare your map, decide what features are important to capture, verify the accuracy of what you captured, and create metadata that describes the captured information.

Forms of digitizing

You can choose from three basic methods of converting from *analog* (hard copy) to digital GIS data. Manual or hand digitizing is perhaps the oldest and still a time-honored, albeit tedious, method. Scanning involves using advanced versions of the technology of the modern digital camera and personal flatbed scanners. Finally, heads-up digitizing is an increasingly common form of input. The following sections give you a brief idea of how each method works, which can help you decide which you're comfortable with.

Hand-digitizing data

Hand digitizing is a bit like hand washing. Some delicate things really do benefit from special attention. Although many operations today have abandoned large-volume hand digitizing, you may find yourself in a situation in which you have a special map or your scanner's inoperable, making an old-style digitizing tablet really helpful. In some instances, hand digitizing has become a lost art, but you may like to remind yourself of the old days — plus, you do have a lot of control over what you enter into your computer when you use this technology.

A digitizer is much like a mouse with an attitude, but the mouse pad also has a role to play. The mouse, in this case, is called a *cursor* or a *puck,* and it normally has a small transparent Lucite portion that sticks out. The puck has a nicely defined crosshair target, so you can see exactly where it's pointing. A board (the tablet or table portion of the digitizer) replaces the cushioned mouse pad, and this board includes a very fine mesh of electrified wires that crisscross up and down, and right and left, at very precise distances. These wires are sensors that know exactly where that target on your cursor is at any given time. And because the board and the cursor talk to each other, the digitizer software can record the position of the cursor in exact millimeters or small portions of an inch.

Really nice large-format digitizers (like the one in Figure 8-6) have a lot of space on which you can place large maps. You can raise and lower the digitizers, and tilt them to accommodate any user.

Figure 8-6:
A large-
format
digitizing
tablet.

A digitizer is much more than hardware, though. It also has software that allows your computer and its GIS software to communicate with the digitizer. The digitizer and the GIS need this software because the coordinate system on the digitizer is flat and measured only in inches and/or millimeters, so it doesn't understand that the map you have is based on a scaled-down, projected version of the real world. (Check out Chapter 2 for more about coordinate systems.)

The digitizer usually has its own graphics language and hardware drivers that either come included with your GIS software or must be installed on your GIS. Not to worry — virtually all large-format digitizers are designed for all forms of graphics communication, including GIS. Most GIS software already contains the drivers for the digitizers on the market.

Scanning your map

Both scanning technology and the software that interprets the results have improved enormously over the past three decades. You may already own or use a simple scanning device, such as a copy machine or small flatbed scanner. Although these types of scanners are a similar technology, for most maps, you want to use a large-format drum scanner (like the one in Figure 8-7).

Figure 8-7:
A drum scanner.

You place the map or image edge-first onto the drum inside, and the drum spins while a laser reads *(scans)* the document. Normally, the scanning software converts the document into a raster image that your GIS software can easily convert to vector format. Earlier drum scanners were very expensive and frequently produced enough errors that data input specialists often found it faster and more effective to digitize most maps by hand.

Using heads-up digitizing

Because much aerial photography and even many maps have already been scanned as graphic images, more GIS data input specialists use those graphic images as a background template against which they do digitizing. You display the image on the computer monitor and then selectively isolate and record the locations of the image features inside the computer (as shown in Figure 8-8). You can most effectively do this type of digitizing when the graphic includes many locations for which you've already well established the accurate latitude and longitude locations.

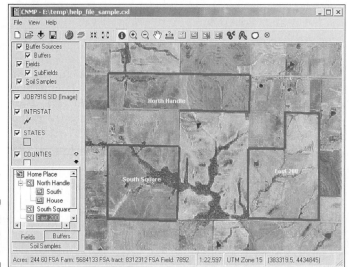

Figure 8-8:
Heads-up
digitizing.

As compared with the other forms of spatial data input, heads-up digitizing is more direct, requires less equipment, and doesn't require you to constantly relocate *tic marks* (little x's that identify the corners) on the maps or aerial photographs that you're inputting. I've also found that I can sit and input considerably more comfortably than I can while standing over a digitizing tablet. Also, when you use heads-up digitizing, you can easily see whether you've included all the features you want because all the features you have digitized appear as graphic lines or points over the background image.

Heads-up digitizing does have some drawbacks. Most of those drawbacks result from the need to scan the raw imagery and maps to make them compatible as digital background for the heads-up digitizing process. With today's scanners, you don't really have to worry about this issue, especially when compared with the variable distortion of the imagery or paper maps themselves, which may occur when digitizing over multiple sessions under uncontrolled temperature and humidity.

Preparing your map for digitizing

When you digitize or scan a map, you have to do several things to prepare the map and yourself for the process. Digitizing some maps takes a very long time, so you can get the most accurate results by digitizing those maps in a series of short sessions because you will be less fatigued while you work.

Follow these steps to get your maps ready for effective digitizing sessions:

1. **Put your maps in the room where the digitizer does its work and leave them for a few days.**

   Doing this lets the maps get used to the temperature and humidity. Yes, the environment does make a difference — recall all those sagging posters that you've seen on walls that used to be stretched taught?

2. **Remove any obstructions or marks from the map.**

   The scanner records everything — including Spot's hair, if it's on the map. You don't need to worry about this concern as much if you're hand digitizing with a tablet, rather than scanning.

3. **Secure the map to the tablet, if you plan to use a digitizing tablet instead of scanning your map.**

   You need to tape the map to the tablet so it doesn't move around. Use drafting tape, not masking tape. Masking tape will tear your map.

4. **Mark the sequences of points, lines, and areas (polygons), that you plan to digitize during a given session.**

   You can save time by picking out and marking only those features that you want to scan. Figure 8-9 shows a taped map with sequences of digitizing locations in place on the map.

If you're really serious about preserving your map, place a tight surface of clear acetate over it and mark the acetate so that the map itself remains pristine.

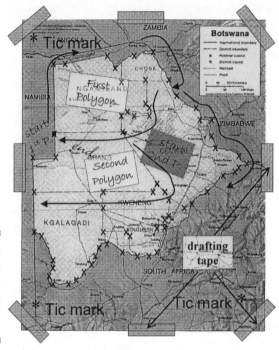

Deciding what to digitize

Only you can figure out what you need to digitize on your map. First, decide what you want out of your GIS, formally called the *spatial information product* (SIP) — meaning what you need your final map to look like. If you know what the final map looks like in your head, you can figure out the ingredients you need to make that map a reality.

This map preparation isn't simple, but it isn't as daunting a task as you might think, either. You need to *decompose* your project — break it into individual bite-sized pieces rather than trying to digitize large complex maps all at one time. Nearly every SIP is a composite of all or pieces of various maps. For example, a map showing the best location to put a new school in your town would be an SIP.

If you ask yourself some questions, you can get a pretty good idea of how a map is created. Suppose that your general question is, "What makes for a good school location?" Here's the list of features a school needs and the related information that a map needs to show:

✔ **A school needs a location near students.** So, you need a map that shows where potential students live.

✔ **A school needs a location that's not already built up.** Okay, you need a map of available land for sale.

✔ **A school needs a location that's accessible and legal.** You want to position a school near streets (so, look for a street map), where it can access utilities (water, sewerage, electricity, gas maps) in an area that's zoned for public buildings such as schools (a zoning map), and so on.

Cleaning up after digitizing

The result of digitizing maps — getting them into your computer — is to create a database for your GIS to use. But during the digitizing process, some inaccuracies may creep in. You need to look for two general types of errors when deciding whether your GIS database is ready for use: *entity* (graphical) and *attribute* (description) errors. Here's a list of general rules that can help you decide whether your database is correct (keep in mind that many of these rules overlap):

✔ All entities that you want to include are present.

✔ Your database contains no unnecessary features.

✔ All entities are in the correct location.

✔ All entities are the correct size and shape.

✔ All objects have the correct labels.

✔ Each polygon has exactly one label.

✔ All features are of the correct type.

To determine whether the database includes all the features you want or whether it contains unnecessary features, you often need a complete accounting of what you wanted to include in the first place. You prepare maps prior to hand digitizing so that you know exactly what you need to digitize and what should be in your database when you are done. Because you can actually see the relationship between what you have completed digitizing and what you have not completed right on your computer monitor, heads-up digitizing has advantages over traditional digitizing methods.

Entity errors that relate to the accuracy of location and position, collectively called dangling nodes (*dangles,* for short), come in two types — undershoots and overshoots (see Figure 8-10). Overshoots and undershoots might cause your GIS to either over-report or under-report the number of polygons in your database, which can also make your labels inaccurate.

Dangling Nodes

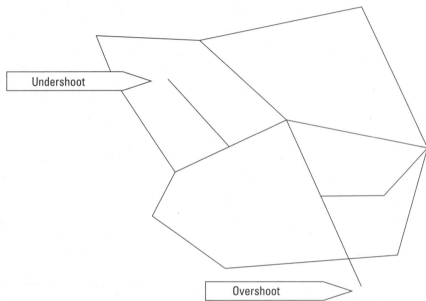

Undershoot

Overshoot

Figure 8-10:
Dangling
nodes, or
dangles.

In most GIS software, digitizing modules ask you to define the nature of the features while you input them by assigning specific codes that indicate, for example, the beginning and ending of a line or polygon. If a line coded as a polygon fails to connect to create that polygon — if the line is a dangle — the software typically displays a graphic symbol indicating there is a problem.

In my experience, you can more easily find and correct overshoots than under-shoots. Unless you're Ms. or Mr. Stable, and you are very certain about exactly where you are digitizing, draw those lines too big, rather than too small.

Building the metadata

Some inexperienced GIS practitioners might lead you to believe that if you have a clean, error-free database, you're ready to store, use, and share your data. Nothing can be further from the truth. I can't tell you how often I've received databases, created with meticulous care, that still don't really explain exactly what they contain, how accurate that data is, the timeliness of the information, what the categories mean, and so forth. I needed something that describes the database. What I needed, and what you need to make your GIS database really useful, is metadata.

Metadata is data about data. It's a complete description of the major facets of quality, source, and accessibility of the data that make them useful not only to you, but to others who might use the data after you do. In my teaching, I've used the phrase "tracking down the wetlands fox" from a newspaper article I saw years ago. That article described the difficulty of sharing databases when some of the layers contained multiple definitions of something as seemingly obvious as a wetland. Because it has different official definitions for different agencies, a wetlands map no longer contains distinct identifiable categories. Comparisons of different wetlands maps become impossible.

You can find many other examples of classification discrepencies, including comparisons of old soil maps with new ones, trying to link land-use patterns and land-cover patterns when the terms used are nearly identical for substantially different classes, and more. But these examples constitute only one type of metadata. The Federal Geographic Data Committee, which is responsible for maintaining spatial data standards in the U.S. Government, has developed a list of seven basic entries for geospatial metadata:

- **Identification:** Defines the database name, developer, area of coverage, categories, themes, collection methods, and data access

- **Data quality:** How accurate and complete the data are, adaptability for application, positional accuracy, completeness, consistency, and sources for derived data

- **Spatial data organization:** Data model (Vector/Raster), number of spatial objects, additional non-coordinate methods of location encoding

- **Spatial reference:** Latitude and longitude coordinates included, projections used, datums, parameters used for coordinate transformations

- **Entity and attribute information:** Specific attributes, what information is encoded, codes, and code definitions

- **Distribution:** Where and from whom to obtain the spatial data (you can often access and read the metadata before you actually obtain it), available formats, media type(s), online availability, and cost

- **Metadata reference:** Compilation date and compiler

There are many excellent examples of metadata. Here is a URL for a typical set from the USGS vegetation database for a portion of South Florida:

```
http://sofia.usgs.gov/metadata/sflwww/vegmap.html
```

People (often, government agencies) have created numerous online tools for developing metadata. You can find information about some of these tools at `http://sco.wisc.edu/wisclinc/metatool`. Today, most commercial-grade GIS software includes such tools for you, so you don't need to use software and interfaces that you don't already know. Moreover, the vendors recognize the need for compliance with established metadata standards, so the better companies always do their best to be both compliant and current. Most GIS software also includes a tool for including metadata with your datasets (Figure 8-11).

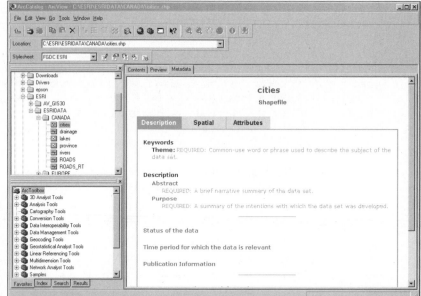

Figure 8-11:
The
metadata
tool inside
ArcGIS.

Always provide the most current, complete, and accurate metadata you can. By the same token, never buy data for your GIS from a vendor who doesn't supply complete metadata.

If your GIS software doesn't include a metadata tool, you may need to find a third-party tool if you plan to share your data with others or your data must comply with data standards. Although such third-party metadata tools exist, you may find that using one is inconvenient. Creating, saving, and linking the metadata is easier when the tool is integrated into your GIS software.

Part III
Retrieving, Counting, and Characterizing Geography

The 5th Wave   By Rich Tennant

"Yes, I know how to query information from the program, but what if I just want to leak it instead?"

In this part . . .

The data you put into your GIS isn't much use to you if you can't search out specific data, compile that data, and then analyze and interpret to make sense of it. This part shows you how to retrieve just the right data and then turn that data into the information that meets your needs. In this part, you find out how you can describe the geographic data you find both numerically and by type. You also see how you can order geographic features by their sizes and shapes, as well as by their locations — both in absolute and in relative terms. Searching out and characterizing geographic data are powerful skills that serve you well when you work with GIS.

Chapter 9

Finding Information in Raster Systems

*G*rid-based maps indicate where geographic features are located on a geographic grid, how much space they occupy, and their locations relative to other features on the map. So, grid-based maps give you an extremely easy to understand way of storing spatial information.

With paper maps, you get the information you need by reading and interpreting the printed map. Paper maps are limited because you can layer only so many geographic features onto one map before that map becomes unreadable. For example, even though each kind of feature relates to transportation information, you can't effectively show highways, waterways, railways, and ATC flight patterns all on one paper map. That would be one confusing map! You'd probably divide this kind of information onto separate maps. To work with more than one or two types of information, you need to gather information from separate maps — and who wants to go through that effort?

In GIS, you can comfortably map many types of information to a geographic grid because all the information can be coded into a computer and selectively retrieved for whatever subset of data you want. You use computer search and retrieval techniques to get just the information you need. The GIS may contain geographic information about all crops grown in the United States. You can retrieve just the geographic information about soybeans in part of Kansas, if that's what you need. First, though, you need software programs that allow you to locate the features you need to find. This chapter gives you the rundown on the easiest, most hassle-free ways to get the information you need from your GIS.

Creating a Search Strategy

When you look for information on a map, you look for geographic objects and distributions that match the theme, or classification system, of the map. You might look for a street name on a road map, a town or state on a political map, a rock formation on a geological map, or a type of plant on a vegetation map.

On one map, you might be trying to find the boundaries of your property or the area that's expected to flood during severe flood events. When you search for this type of information, you're looking for point and line objects, as well as areas and the boundaries around those areas.

The level of geographic data measurement (nominal, ordinal, interval, ratio, and scalar levels) determines the mathematical nature of the data turned up by your search. For more information about geographic data measurement levels, see Chapter 3.

After you determine the topical, classification, and measurement constraints for a search, you're ready to choose an analysis technique. GIS offers many complex analysis techniques, which I discuss in Part IV of this book.

An example of a complex analysis might include combining maps to determine the best place to put a dam based on the amount of water in a stream, the shape and size of the river valley, the number of people and homes that might be impacted, and the cost of putting the dam in at that location. The number of factors involved in analyzing where to put a dam requires you to be able to extract all sorts of information about the stream hydrology, the terrain, the surrounding population, and economic costs involved under certain conditions.

All such analyses require software that can find the geographic features used in the database. Complex analysis often drives the search for information. Complex approaches nearly always require fairly simple forms of search strategy. So, whether you're interested in complex analysis and modeling, or the plain vanilla results of a search, you need to understand the search process.

Locating objects on a map

When you read a map, you often need to find objects that you know exist somewhere in the map — you're just not sure exactly where they're located. In grid GIS, if you know the names, categories, or measurement levels of the object, you ask the software to find all the grid cells that share the name, class, or data level you seek. The software finds objects by the codes you use to identify them when you create the data. After you decide on a specific search object, you can follow these general steps to find the object and verify the results in most raster GIS:

1. **Select the map layer that contains your search object.**

   To locate all public school buildings, for example, you first select the map layer that corresponds to the object *schools*. You wouldn't look in a map layer called *hydrology, soils,* or *elevation* to find the location of the schools. But a map layer called *public buildings* probably contains grid cells categorized as courthouse, library, civic center, and many others — including schools.

2. **Specify the names, codes, measurements, and so on that identify your search object.**

   Look in the data that you or someone else produced to find out the code used for each object when the grid layer was created. In this example, you would determine the code used for schools and tell the computer to search for that in the grid layer called *public buildings*.

3. **Review the retrieved information and make sure you got what you searched for.**

   The final map contains a subset of the data in the public buildings layer and will highlight only the cells containing the code for school buildings.

4. **Name your new map.**

   You get the honor of providing a new name for the map. In this case, you'd likely call the map *schools*.

Folks run searches for all sorts of reasons, and the most simple (but not necessarily most common) is curiosity. You might want to know how the schools are distributed throughout the city. Or maybe you just want to simplify the map so that you can easily see individual schools without having to deal with all that annoying background. Most often, though, when you search for objects, you have a more complex question in mind. You may want to search for schools because you want to know where they're located relative to other features, such as housing, so that you can buy a home near a public school.

People almost always have features or distributions other than simple location in mind when they perform a search in GIS. For example, you might look on a map for the location of your home, but you really want to find out whether you live within the same school district as a particular elementary school you want your children to attend.

Searching for linear features

Linear features are common search items. Among the most common linear features are roads and other transportation routes, such as rail lines. In grid GIS, these features appear as a series of connected grid cells within a transportation layer. Within that layer, you might be looking for just the Interstate Highway system in the United States or the Great Siberian Railroad within Russia.

You may be searching for these features to find the shortest route from one place to another, which requires the software to determine which roads would suit that search. Or you may want to know which roads allow hazardous cargo so that you can perform a later analysis of populations at risk. The software finds these linear features the same way it finds point objects — by the codes you use to identify them.

Searching for areas and distributions

You may want to search for areas in a GIS, such as large cities, counties, states, farmlands, or industrial areas. Or perhaps you want to determine distributions of birds and other wildlife, vegetation, disease, religions, population types, natural resource deposits, or literally thousands of other features. In fact, Roger Tomlinson developed the first GIS mainly to map resource distributions.

Searching for *polygons* (areas and distributions) in a GIS helps you determine how much of your specified search subject exists (which can be important information), but your search results can provide other insights, as well. For example, you may want to know if

- **Distributions are growing or shrinking.** You may need to know the location of the current distribution of the imported fire ant or the Africanized bee in the United States and how that distribution changes over time.

- **Changes in distribution might affect other point, line, and area features.** With information on the scope and movement of exotic (and often troublesome) species, you could prepare for their migration into areas that they don't currently occupy.

- **Existing conditions might affect your future plans.** You may want to know the locations of areas zoned for business so that you can make decisions about where to start a new business.

The example in the first two bullets of the preceding list show that time can be a factor in your search. So, if time is coded into your database through layers derived from different time periods, you can search for locations and distributions that satisfy the time attribute code just as easily as any other code.

In a grid-based GIS, areas are comprised of groups of grid cells that share common code numbers. So, like with point and line features, the software begins by searching in the appropriate layer. Then, it identifies the groups of grid cells based on how you formulate your search for the existing codes.

Using the Software to Perform a Search

In raster GIS, you most often search by actually finding the categories within each layer rather than identifying the features as objects (for example, polygons or lines). You can search with your software's query tools by making selections that pick out categories in each grid layer. And so, you most often search for the attributes contained in each layer by using a menu system rather than by finding the objects on the map itself.

You can also search for objects (for example, a group of cells representing a category you're interested in) on the map by pointing at the cells themselves. But when you do, your search retrieves all the data for those cells and not just the particular category data you may be looking for.

Searching in simple raster systems

In simple raster systems, you look for grid cells to find collections of cells representing points, lines, or areas. The type of search strategy involves selecting the categories that you want to work with within each layer. This selection process is simple, and you often make selections by using a database query tool like the pull-down menu system shown in Figure 9-1.

The graphical search allows you to define a focused study area (sometimes called an area of interest) right on the map, and it captures the features associated within other map layers and produces a new layer. When you select an area of interest, the software doesn't separate out different categories, but instead picks up everything that is contained in your area of interest. If you want to find individual categories by using the map, you can point to grid cells within each area you see (check out Figure 9-2), and the GIS provides you with the grid cell's location and descriptive information about the category it represents.

After you retrieve data from your search, you can decide to rename or reclassify it using the same user interface you used to locate it in the first place (see Figure 9-3). This multi-purpose interface gives you the power to pick and choose what you want to see and control how it's grouped together.

Attribute searches in simple grid-based systems find the codes associated with each category or class of data directly. Because no DBMS tables are associated with these grids, the search is simple and often obvious because the categories and classes of data for each map show up in the legend. So by looking at the legend, you can easily see all the information that's available (none of the attributes are hidden).

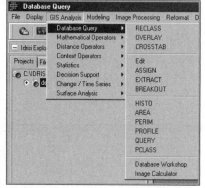

Figure 9-1:
A database
query
tool with
pull-down
menus.

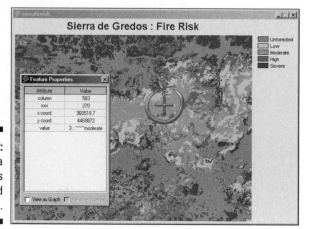

Figure 9-2:
Getting a
grid cell's
location and
description.

Figure 9-3:
You can
reclassify
data you
retrieve
in GIS.

Searching DBMS-supported raster systems

Raster systems that have associated database management systems (DBMS) allow the same approach as vector systems to finding points, lines, and polygons. You can still perform searches based on spatial query, graphic, dimension, data type, and attributes. You can do dimensional searches in these more advanced grid-based systems because they have an associated DBMS, which you need for this approach.

The primary difference between the database supported type of system and the older, more basic grid-based systems is that the software can store and manage far more complex and detailed categories. The attributes aren't limited by what the map legend can easily display. So, the attribute search is much more powerful with these systems than with their earlier counterparts.

For example, say that you have a forest map in a simple grid system. It contains a set of forest types, each of which appears in the map legend. The more advanced DBMS-supported system could have a set of supporting attribute information (such as age structure, fire frequency, fuel buildup, disease or infestation condition, treatment, estimated board feet, and so on) associated with each forest type. This association removes the need to have a new map for each of these categories. And now you can search for pine trees (a basic category in your forest map) that are known to have pine park beetle infestation and haven't been treated, which saves you from having to perform three separate searches and then combine them.

Counting and Tabulating the Search Results

After you find objects and distributions by using your available query language or interface, you want to get some numbers associated with those objects and distributions — and fast. This tabulation allows you to count numbers of objects, lengths of road, areas of distributions, and even volumes of land. To count the number of point objects, you need only count the number of grid cells. I deal more with measuring geographic objects and distributions in Chapter 11 and 12, but many of these basic counts are displayed without any need for advanced searches that you might want to perform later if you are interested in tabulating complex measurements.

So, with these advanced searches, you can also tabulate and summarize sizes, lengths, areas, circumferences, and other types of measurable data. In fact, most grid-based systems often report some of these numbers immediately upon completing a search.

Getting simple statistics

Among the most powerful basic tools of the GIS is its ability to tabulate the numbers you collect and provide you with descriptive statistics. These descriptions include maximum and minimum values for a search, measures of central tendency (the mean, median, and mode), and measures of dispersion or difference around the measure of central tendency (e.g. standard deviation).

So, beyond collecting and tabulating the number of towns that you find in a search, you can determine which of the towns is the largest and which is the smallest. If you have this information, you know the range of sizes (a measure of dispersion) and can rearrange, compare, and otherwise analyze your results. You can

- Sort these statistics or group them by size categories, as well.

- Derive basic statistics — such as mean, median, or mode — that give comparative measures with larger and smaller towns.

- Look at measures of dispersion — such as standard deviation or other statistical measures — that show how much variation exists from the measure of central tendency.

You don't need to know the details of basic descriptive statistics, but professional GIS software typically offers basic, as well as more advanced statistical techniques. Of course, you can use these techniques only if you have attribute data for the entities represented by the grid cells you select.

Interpreting the results

After you collect, tabulate, and prepare summary statistics for your data, you get to decide what you want to do with the information. Descriptive statistics communicate what the data tell you in such a way that you don't have to be a GIS expert, or even a map reading expert, to understand.

You can often most usefully present the results of your queries and summations by using graphs and charts. When you include graphs and charts, they may be most effective when you present them right on the map itself (but not all software does this automatically). The combination of graphs and charts with an output map (see Figure 9-4) allows you to point to features on the map that are associated with the summaries.

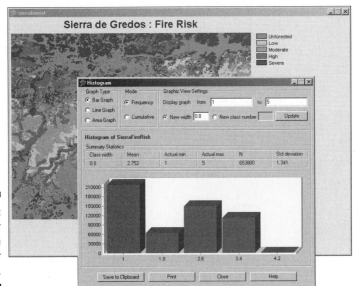

Figure 9-4:
A bar
graph from
a raster
GIS layer.

When displayed next to your map output, or on the map itself, the statistics have a geographic component. Big features aren't just big — they're big in a particular place, which communicates the importance of space and place to your audience. In other words, this type of presentation can often enhance your ability to communicate all your ideas. And the better able you are to communicate, the more effective you are in your position. So, GIS makes you look good.

Chapter 10

Finding Features in Vector Systems

· ·

· ·

The most graphically accurate method of representing maps in a GIS is the *vector* system (which I describe in Chapter 4). Vector systems represent geographic space by using a series of points, lines, and polygons. This method differs from *raster* GIS, in which geographic space is stored as a collection of squares called grid cells. Because data storage differs between raster and vector systems, the way you search for those data also differs.

Not to worry. In some ways, how you find geographic objects in a vector GIS is more like how you find them on paper maps. Vector GIS allows you to work with geographic features in a more realistic way than you can in raster systems because the points, lines, and polygons are more realistically located and correctly sized. For example

- ✔ **Point objects:** Such as wells and houses display like they don't take up huge chunks of space (or an entire grid cell).

- ✔ **Lines:** Have the appropriate lengths but aren't so thick that they're unrealistic.

- ✔ **Polygons:** Represent the boundaries of areas much more exactly than grid cells and don't have that blocky, checkerboard appearance.

Some books include lists of advantages and disadvantages of vector over raster GIS, but I prefer to think of these items simply as differences. Vector GIS is different in the ways it stores data (see Chapter 4), lets you retrieve data, and displays the results of data analysis. In this chapter, I tell you about the retrieval of data, and I concentrate on the differences only in the way that they affect retrieval.

Getting Explicit with Vector Data

On paper maps, you read the symbols that represent features to identify those features visually. In vector GIS, you use the power of the computer to identify the stored features by name or description. You usually need to find features on maps so that you can use them in more complex tasks, such as finding routes from place to place, comparing different maps, and finding the best place to put a business. (GIS can lead you to business success by helping you evaluate potential markets. I talk about some of these exciting applications in Chapter 1.)

Unlike raster systems, vector GIS systems have location information that's more accurate and visually more like the real world, so you can search geographic space to achieve more accurate results. You can search by looking at the tabular information or by selecting the graphics on the screen. Typically, the software has its own graphical user interface (GUI) that enables you to do these searches (see Figure 10-1).

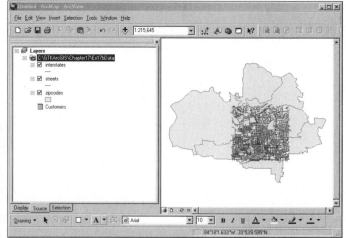

Figure 10-1:
The ESRI
ArcGIS
vector
GIS user
interface.

From your perspective, this approach to finding features might appear pretty much the same as in raster systems because you don't really need to know what's going on inside the computer. The differences show up in certain types of analysis — particularly those involving measurement of distance, direction, perimeter, area, and so on — and are often an automatic result of the data retrieval process.

Seeing How Data Structure Affects Retrieval

Like grid-based GIS (those that divide geographic space into little squares called grid cells), polygon-based GIS also comes in different flavors. These flavors are often not so much a function of the user interface, but rather how the computer goes about retrieving the data you ask it to retrieve. The three major types of systems are network, relational database management system (RDBMS), and object-oriented GIS (see Chapter 6 for details on these systems). In the following list, I describe the two systems that you're most likely to encounter in your GIS work (I purposely ignore the network systems because they're pretty rare):

✔ **DBMS-based vector GIS:** In this traditional structure, you use some on-screen graphics to pick out features for retrieval, or you use a search engine query to find the features' attributes. The graphics link to the tabular data containing the attributes in the DBMS, and the two elements (graphics and tables) work together to supply you with data that satisfies your search. When you find the attributes, you get the associated graphics, and when you find the graphics, you also pull up the tabular data. Your searches enable you to combine data and link features together in any way you want so that you can create new maps.

One example of a DBMS-based GIS is Intergraph's GeoMedia package. Emergency services and military applications often use this software package to plan routes and determine the placement of troops and equipment.

✔ **Object-oriented vector GIS:** For object-oriented systems, you use pretty much the same search strategy you would in the traditional DBMS system. But the object-oriented system is somewhat smarter about how it retrieves the data. In an object-oriented system, each object represents a geographic feature and is associated with not only descriptive attributes, but also a set of rules about how the objects relate to each other and to their environment. In object-oriented systems, such properties (the attributes) and methods (the rules) belong to a grouping of objects called a *class*. Any object that's part of the class group is said to *inherit* these properties and methods. In short, the rules prevent you from making mistakes by retrieving and linking incompatible objects.

Here's a classic illustration of how object-oriented systems can assist you in retrieving valid and useful information. Suppose you're the Director of Transportation Planning for the City of Barstow, and you have a new development coming in. You know this new development needs roads, and you include the following interaction with the GIS in your planning process:

1. **You establish criteria and assign an assistant to create a proposed set of roads and streets for the new subdivision inside the GIS.**

   The network of new streets must be able to handle a certain flow of traffic (number of cars), must be accessible by all the future homeowners, and must comply with the usual traffic rules. You also remind the assistant that this road network must link to the existing roads.

2. **Your assistant searches the map of the development area with queries designed to match all the criteria for new roads to be placed in the new subdivision.**

   Here the assistant tries to find where links to existing roads should be, where roads need to connect to new subdivisions, and what traffic laws might exist. Then the assistant produces a new map, and adds additional roads that are meant to satisfy these criteria, including a few one-way streets that allow traffic to flow from north to south only.

3. **You look at the assistant's map and review the new street pattern.**

   Here's what you find in the new development's street map:

   - **From a GIS with a traditional RDBMS structure:** You point out that the new one-way streets (with traffic flow from north to south) are connected to one-way streets that flow from south to north in the developed part of the city. Because the traditional DBMS does not contain rules that prevent such mistakes, your assistant has been able to place streets with connecting roads that flow in opposite directions.

   - **From an object-oriented GIS structure:** You see street patterns showing traffic flow that follows the proper direction of one-way streets. If your assistant had used an object-oriented structure, no matter how hard he or she tried, the system would not have allowed the two opposite street patterns (north to south and south to north) to merge — because doing so violates the rules contained in the objects themselves (the streets). Figure 10-2 shows a traffic route that skirts a bad road connection on the way to its destination.

If you're often concerned about how certain data types that you retrieve will work together, choose a GIS package that has the object-oriented data model. This model's structure validates the proper relationship of the data during retrieval and helps ensure that your combinations work together to produce legitimate results.

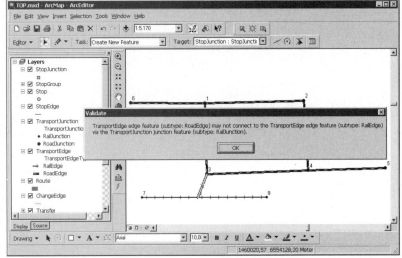

Figure 10-2:
The ESRI
geodata-
base just
says no to
improper
connec-
tions.

Deciding How to Search the Systems

You can search for information in a GIS in many ways: You can grab a set of graphics from the screen, search by neighborhood characteristics and adjacency, or look for a particular data type. Because this chapter discusses the vector data model, I focus on the most important ways that you typically search for geographic objects in a vector GIS, as follows:

- ✔ **Search by attributes, using the SQL interface.** You can search for any attribute that you want, as long as the attribute is included in the database. For example, you can search a forest layer for areas infested with Dutch elm disease or the pine bark beetle. You can search a census layer for census tracts containing households that have a particular level of family income. Or you can search a transportation map for dirt roads. Basically, you can search any characteristic — size, color, variety, value, category, length, area, condition . . . you name it — that has an associated attribute stored in the database (or that the software can calculate from the database). If the information is in your database, the GIS can find it.

  SQL (Structured Query Language) is the computer language used to formulate database queries based on logical or mathematical search operators. Polygon-based GIS packages rely heavily on this form of search strategy, even more than the grid-based systems do. Check out Chapter 9 for more about SQL.

- ✔ **Search for features by using the graphics interface.** For example, you might see on-screen a set of features that you want to work with. You can use your cursor to draw a box around those features to select them.

Targeting the right data source

You need to identify three search criteria before you start a search (in either a raster or vector GIS) to help you make sure that you're looking in the right place:

- ✔ **Theme or topic:** The search for individual features is first directed to the layer that contains data for that theme. For example, a search for geological features such as mesas requires you to use the geology layer.

- ✔ **Classification system:** Knowing what classification systems identify the information you need can help you decide what specific classes to search. If, for example, you're looking for land to use for commercial development, you can find that information in a layer that uses the zoning classification system.

- ✔ **Level of measurement:** The level of measurement determines what search values can find the data you want. For example, if you want to search for land-use codes (nominal data stored as integers), you probably don't want to search for values of 9.3044211 (interval or ratio data) because the codes are not rational numbers.

Keeping the expected result in mind

In Chapter 9, I talk about how a search can drive analysis. Briefly, this concept involves planning your search so that you target the right data source (as discussed in the preceding section), look for the right criteria, and properly analyze the result. (See Chapter 9 for more explanation of the process.)

Follow these steps to work within the search-drives-analysis concept:

1. **Define your question in geographic terms.**

2. **Decide on the spatial features that you need to work with.**

3. **Formulate your search based on your data needs.**

4. **Conduct your search.**

5. **Prepare to analyze the results.**

Locating Specific Features with SQL

You'll probably often want to search for specific point, line, or area features with your GIS. You might want to find specific types of trees (such as oaks) in a layer of trees, or you might want to find a set of slopes that range from 15 to 45 degrees in an elevation layer. The theme (for example, forests or elevation) directs you to the right layer, the classification limits what categories (such as oaks or slope) the software can effectively search, and the level of measurement determines the types of numerical data you find.

In the following sections, I outline how to use vector GIS to find specific features by using the SQL search engine. Each attribute search you perform — whether for points, lines, or polygons — follows the same procedure. I use ArcGIS for the examples, but any vector GIS should work pretty much the same way.

Getting to the point(s)

Perhaps you're interested in the population statistics for the African continent, so you use the GIS data you just happen to have (okay, the data I happen to have) that shows the countries and cities of Africa. The data includes statistics on population both for countries and cities, so you know you have the right data source. You first want to know a little about the cities of Africa, and you start the search process to look for point features (the cities).

Follow these basic steps to perform any attribute-based search by using the SQL tool of your GIS:

1. **Open your database.**

   I load up my GIS and, using the ArcGIS user interface, pull up Africa on my screen. The data layers are active (which means they automatically display on-screen). In Figure 10-3, you can see that the countries and the major cities are already appearing. I can tell that the geographic features I want to examine are already available.

2. **Highlight the theme you want to search.**

   In this example, I want to look at the city data, so I highlight the City theme.

3. **(Optional) Open the attribute table for that theme to examine the data.**

   This process step is optional, but it helps you double check that the tables really contain data or determine what data you might want to search. Use it with your GIS searches — if you're choosy like me and want to know for sure that the data exist. This lets you spread your wings a bit and let the GIS software help you sort through data to see what's there.

   In ArcGIS, I right-click the theme so that the table containing the attributes appears. I see some pretty impressive sets of numbers there (see Figure 10-4). Your software may have a different way of viewing the tabular data.

4. Open the query tool.

The search tool within ArcGIS has its own built-in SQL language. I open the search tool by choosing Selection⇨Select by Attributes. This cool tool, shown in Figure 10-5, does two things that are pretty standard for GIS SQL tools:

- It lists all the categories of data in the database (so that you don't have to know all the categories).

- It enables users to formulate a search strategy based on their criteria.

5. Formulate the query.

In this step, SQL jumps in to help. The interface you use may have drop-down lists to select from, check boxes to select, and so on. In this instance, I want to see only those African cities that have a population I specify. When I select the criteria, the ArcGIS software actually reproduces what I type and puts it in a little box at the bottom of my screen.

6. Apply the query.

Your software may have a button to click to apply the query. In the ArcGIS software, I simply click the Apply button to apply the query. And after I do, the cities I selected based on my criteria are highlighted in the table, as shown in Figure 10-6.

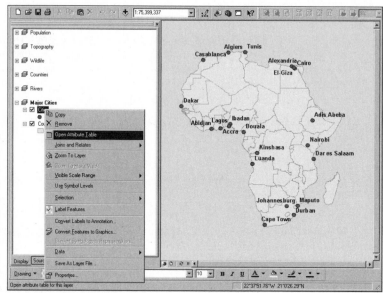

Figure 10-3:
This map
of Africa
shows the
country
outlines and
the major
cities as
points.

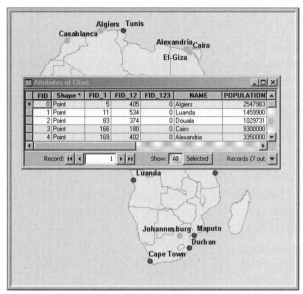

Figure 10-4:
A table of
attributes in
ArcGIS.

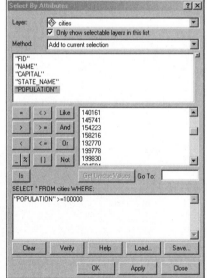

Figure 10-5:
The SQL
tool inside
ArcGIS lets
you pick
your search
criteria.

7. View the results.

Your GIS may show the query results on the map immediately after you
apply the query. Or you may have to do an extra step, such as perform
a map retrieval or map display operation, to see them. After I apply the

query in ArcGIS, I can remove the table from view by either moving it aside or clicking the Close button. Then, all the cities I selected appear highlighted on the map, as shown in Figure 10-7. Because the map and the table are connected, what I select and highlight in the table is also selected and highlighted in the map.

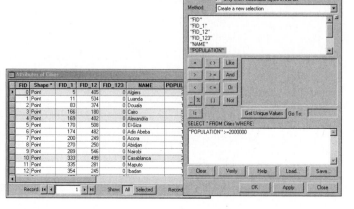

Figure 10-6:
The cities that fit the criteria are now highlighted in the table.

Figure 10-7:
The cities highlighted in the table are also highlighted in the map.

You can perform this type of search for other features (such as lines and polygons) in the GIS, as well. With the database from the preceding example, you might do a quick analysis of country (rather than city) populations for African nations. After clearing all the point search selections, you could follow pretty much the same process to search for polygons (the countries) as in the example with points (the cities).

Suppose you want to see all countries in Africa that have a population of 8,000,000 or more. In the ArcGIS, you highlight the Countries theme (selecting the map from which you want to search for your population attributes), and then open your table. Next, use the selection or search tool to create a query that searches the Countries theme for countries that have a population of 8,000,000 or more. Apply the search. The countries satisfying your criteria are highlighted on the table, and the map also highlights the selected countries (see Figure 10-8).

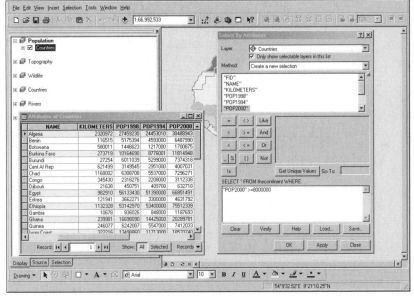

Figure 10-8: Countries are highlighted both in the table and on the map.

Keeping your searches

You can create some really cool layers just by searching for data and then, displaying what you find. And, you save and copy them in case you want to use them again. Each software package works a bit differently, but all enable you to save to the database the layers you create with your searches. For example, some programs use a Save button that has a picture (such as an image of a floppy disk); others use drop-down lists; and so on.

To copy the layers you create in the ArcGIS software, follow these steps:

1. **Find your layer in the list on the left of the screen.**

2. **Right-click the layer name.**

   A pop-up menu appears. One of the menu options enables you to save the layer as a new layer (a copy, if you like). Here's where you have some control over your ability to find the right layers in the future. If you call your creation Mylayer, that might be an accurate name, but it doesn't really tell you much about the layer you created. What happens after you create ten layers and name them Mylayer01, Mylayer02, up to Mylayer10? The numbers don't even hint at the layers' content, and so the names don't help you find the right layer when you return next week to review your work.

3. **Select Save as Layer from the pop-up menu, select a folder to hold your layer, and type a descriptive name when prompted.**

Developing a naming convention for your layers, even the intermediate steps you'll eventually get rid of, can help you keep track of them. Make the names descriptive of either the layer's thematic content (for example, *oaktrees* developed from a search of a layer called *trees*), or the its potential use (such as *interimslopes* for a data layer of all slopes between 15 and 45 degrees that you'll tweak and eventually rename as a new layer called *goodslopes*).

What's my line?

Although both raster and vector systems can represent lines, the vector data models are much better at displaying lines' one-dimensional nature. More important to you, lines on maps represent features such as transportation networks. These networks allow you to model the movement of transportation vehicles, such as cars and trains. Companies use routing programs to plan routes for the delivery of mail and packages. GIS can track GPS-enabled UPS and FedEx trucks along their routes and efficiently direct nearby trucks to on-demand pickup locations. In fact, most goods and services will eventually travel along some form of ground transportation network that's determined or monitored by some form of GIS.

Also, GIS helps with other important tasks related to entities that show up as lines on maps. For example

✔ Finding the fastest route for a firetruck to get to a fire

✔ Determining where a bus needs to go to pick up the most students

✔ Finding out the amount of water flowing through sections of a river or stream (another entity easily represented by a line on a map)

✔ Estimating how much and what kinds of vegetation might occur near a river or stream as determined by the amount of water available.

✔ Evaluating the size of a waterway to determine whether boats can navigate the river

Before you can model transportation or examine the ecology of streams, you need to be able to retrieve and access the pieces of the network and the tabular data that are linked to those features. The selection and retrieval process is easy. In fact, the preceding sections show you how to retrieve population data (a human-related feature) by using points (African cities) and polygons (African countries). Follow these steps to retrieve line data, rather than point or polygon data:

1. **Open your database.**

   For my search, I use the map of Africa and its associated attributes (which I discuss in the preceding sections).

2. **Highlight the theme that you want to search.**

   I want to know which of the many rivers in Africa are navigable because I want to move some equipment along the rivers for my scientist friends. To find the navigable rivers, I choose the theme called Rivers.

3. **(Optional) Open the table for that theme to examine the data.**

   I might take a peek at the tables to make sure the attribute I'm looking for is in them. In this example, the attribute I need is Navigable.

4. **Open the query tool.**

5. **Formulate the query.**

   I use my SQL search engine from the selection tool menu and create a search that looks for all rivers that are listed as Navigable.

6. **Apply the query by clicking the Search button and using the Select by Attribute menu.**

7. **View the results.**

   The table highlights the features I want. In my example, the navigable rivers show up as thicker lines and correspond to the highlighted rows in the table (Figure 10-9).

You can find all sorts of points, lines, and polygons just by letting the SQL search engine rummage through your tables to find just what you want. The location of your study area doesn't matter. As long as you have a GIS system that has points, lines, and polygons — and the correct associated attributes — you can find the data you're looking for.

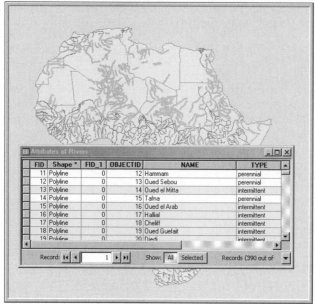

Figure 10-9:
The high-
lighted
navigable
rivers both
in the tables
and the
maps.

Searching Vector Systems using Geography

In Chapter 9, I describe some successful strategies for finding geographic features and distributions. Two of those methods — graphic- and location-based searches — are often combined in vector GIS.

You might assume that the primary purpose of GIS's computer map display is for output, but making that assumption does a huge disservice both to the software and to you. If, as the examples in the preceding sections demonstrate, the tabular data that you used SQL to retrieve are explicitly linked to the graphic display, then you must be able to search the tables by starting with the graphic.

I can use the graphic interface to find things in my favorite Africa database (which I use in the preceding sections). Suppose I want to study the African countries for which I have data on major cities. Visually, I can see these countries because each country that has a major city is visible. But I want to create a map that highlights those countries, and I want tabular data for some later analysis.

The process for finding information spatially in the ArcGIS is actually a bit simpler than using the search-by-attributes method. Because the graphical interface offers many selection options, you don't need to select a layer. Remember, your system may have a different user interface with different menus and buttons, but the concept is the same. Follow these steps to search your data using graphics:

1. **Choose the Select by Location tool or menu.**

   In my example, I chose Selection➪Select by Location.

2. **Select the relative location you want to search.**

   For example, select Adjacent to, Within a Distance of, and so on.

   In my example, I'm going to look for all countries that have data for major cities. To do this graphically, I select the Contain option, so the search will find the country polygons that have data points in them associated with large cities.

3. **Run the Select by Location tool.**

4. **View the result.**

   The software finds all the polygons that have major cities in them and highlights them on the map. And because the map is linked to the graphic, the data in the tables are also selected. Depending on your software, you may have to choose from a menu or click a button to toggle the map or the table on or off.

5. **Repeat these steps to perform another search.**

You can find more about search strategies in Chapter 9.

Counting, Tabulation, and Summary Statistics

This chapter contains several examples of how to search vector systems to find the features and distributions you're looking for. In each case, you have the choice to display the results as a map, a table, or both. The nice thing about retrieving table data is that you can manipulate this data like you can in any database. You can sort, group, and perform calculations on the table data by selecting the available options built right into your software, much like your favorite spreadsheet. Usually, the default settings are quite adequate for the first look, but when you begin to ask more questions, more manipulations allow you to get a better idea of the scope and meaning of your selected data. The following list gives you some examples:

✔ **Counting:** The tables allow you to count the number of items in each column so that you know the total number of items included in the table. For example, I can count the number of countries in my Africa database.

✔ **Tabulation:** Like you can in spreadsheet programs, you can sort the entire table contents by any column. And you can sort each column in either ascending or descending fashion. Suppose that you want to list African cities from highest to lowest population. You can sort the entire table or just the content that your search selected. By simply rearranging the content — without any numerical manipulation — you can examine your data in any order you want and compare one column against another. In many cases, this comparison leads to even more ideas for future comparisons and analysis.

✔ **Summary statistics:** A natural extension of counting and tabulating data is the calculation of summary statistics. Most GIS software allows you to easily create summaries. With a button click or two, the ArcGIS options provide the basic summary statistics — minimum, maximum, sum, mean, and standard deviation. These numbers cover all the basic stats you need to decide whether you may find the numbers useful for further analysis. You can

- Look at the relationships between highest- and lowest-populated countries.

- Determine whether the mean is a good summary statistic for national population in Africa.

- Compare each nation's population to the total for the entire continent.

And here's the best part: You hardly have to do anything to get these numbers.

Validating the Results

When you search for something in a GIS, you most often have some idea of what you want to do with it. The results from a simple search are rarely the same as they would be for a detailed analysis. Still, you can do a few things when you finish a search to interpret whether you did your job correctly. Ask yourself these questions:

✔ **Did I get the correct data?** This doesn't mean did you get roads when you wanted rivers. If you search for rivers, but only navigable ones, make sure the rivers that show up are coded as Navigable. Choosing the wrong attribute can happen, and that causes mistakes, so check that your results returned what you think you asked for.

Getting the correct data also assumes that the data source contains the right level of measurement and a classification system that works for your application. Getting a soil map of Kansas only to find that the map is based on a Russian soil classification system doesn't do you much good if you are looking for a soil in the US classification system.

✔ **Are your data on the right scale?** If you're interested in road networks for a city but search on a database for an entire country, you face a good chance that the smaller roads don't appear in the database. Usually, this situation occurs when you don't have access to the level of detail you want. Be prepared to throw stuff out if your results are inaccurate.

✔ **Are the data timely?** If your data source has information from different time periods, double check that you're asking for results from the right one. Looking at the population of African countries by using the 1998 data doesn't return results from the census conducted in 2000.

✔ **Are the data complete?** Having complete data means that nothing is missing (duh), but it also means that your data offer more than just the database objects. You need to know exactly what the categories mean, where the data came from, and where you can go to get updated data. This data-about-data is called *metadata,* and it functions to provide additional information about a database component's origin, format, synopsis, and so on. Most GIS include a way to add metadata (as I describe in Chapter 8), but sometimes GIS databases don't include the metadata. This omission limits the utility of the data.

Chapter 11

Searching for Geographic Objects, Distributions, and Groups

*A*fter you put geographic data into the computer, you use GIS frequently to get information derived from the data back out. To retrieve data that best suits your needs, you must know why, where, and how you search for geographic objects (such as roads, rivers, lakes, power plants, and thousands of others) and distributions represented by polygons (such as the range of an animal species or ethnic groups within a nation).

The vast majority of geographic objects that you encounter every day occur in groups. Some of these groups even have specific names. Birds occur in flocks, cattle in herds, trees in forests or orchards, quail in coveys, geese in gaggles, and the list goes on. Beyond the biological groups, glacial features (called *drumlins*) form in swarms, streets form networks, streams form watersheds, homes form neighborhoods, and many more. People assign special names to groups to provide a handy way to refer to features that look alike, act alike, occur in similar places, are near certain features, or occur in similar patterns or densities.

In this chapter, I show you how to specify search criteria that can yield the results you're looking for. And after a brief look at common GIS interfaces, I explain the six methods for searching for polygons and why you might need to use each one. You can search for groups of objects by using the same GIS functions that you use to search for individual objects. Then, you can classify the whole group to suit your needs. This chapter shows you how to conduct a search for groups of objects, too.

Searching Polygons in a GIS

When GIS first came on the scene, resource specialists mainly used it to examine large portions of natural resources that were distributed over large areas of land. Similar data that spreads over large areas, represented by polygons (in vector GIS), are called a *distribution*. Distributions are still among the most important features in GIS. GIS can represent distributions of people, animals, disease, crime, land use, pollution, and more. To get the most from GIS, get familiar with distributions and use that information to make decisions. (See Chapter 3 for more about distributions.)

Whether polygons represent distributions of individual objects (such as people or animals) or area features (such as land parcels or forests), you can use many of their properties for later analysis. The data inside polygons can exist at any of the measurement scales, including nominal, ordinal, interval, ratio, or scalar. (I talk about measurement in Chapters 1 and 3.) You can look for objects based on thematic content, such as types of animals, houses, or trees. You can describe distributions based on size, orientation, perimeter, area, change over time, and many other factors.

Polygons are connected to each other, surround other things, and are surrounded by other things. They have a long list of spatial relationships that other objects don't have. To get the most from GIS, you use those properties and relationships to make decisions regarding the data you retrieve when you search your GIS.

Searching for the Right Objects

When you search for an object, you must know what you need to accomplish with it before you can determine which object can do the trick. You look for your car keys because you want to drive your car, you look for your receipts because you need to do your taxes, and you look up a telephone number because you need to make a phone call. You have the same basic reasons for searching for two-dimensional objects (such as land parcels, landforms, or soil polygons and their distributions) in GIS.

You can decide to conduct a search for a bunch of **reasons**. You may want to search for objects within a polygon out of simple curiosity. You may want to open a map just to see the map itself and identify any patterns that appear. Even a quick perusal of map content shows you

✔ **What categories are available:** For example, what land-use categories are present in a land-use map.

✔ **The general size and shape of the polygons:** For example, where pine bark beetle infestation is occurring in a forest stand.

✔ **The general configuration and locations of polygons:** For example, where populations at risk from potential hazardous materials spills are distributed.

Think of searching a polygon as taking a journey through its data to discover what's where. Think of yourself as a world explorer and think about your maps as questions, rather than answers. When you get in this mindset, such questions (or *hypotheses*) lead to additional exploration: searches, measurements, and subsequent analyses.

Extracting specific information

Simple curiosity is fine, but a more common reason you conduct a search of polygonal objects and distributions is to extract specific information from the maps. You may conduct a search to simply explore the data, but most often, you use more complex analysis after data exploration or as a result of previous knowledge of the map content.

You gain knowledge about map content by viewing similar maps or by drawing on your familiarity with the real world represented by the map. For example, you might examine the locations of different land-use types relative to different elevation zones. You can examine the relationship of your maps' features by looking at the maps, their legends, and any associated tabular data (for example, a road map might have a table showing the mileage between major cities).

Searching for objects in modern GIS requires a logic-based search strategy, which a graphical user interface (GUI) provides (as shown in Figure 11-1). The GUI uses Structured Query Language (SQL), which includes basic AND, OR, and NOT operators, as well as mathematical comparative operators such as =, >=, <, and many others.

Misplaced parentheses or other punctuation marks can wreak havoc with searches that use SQL search methods such as the one shown in Figure 11-1. Until you get really comfortable with the syntax, do searches in small bites, creating interim maps that you analyze, combine, modify, and otherwise work with later. This approach forces you to think clearly about what data you're looking for and formulate your queries succinctly.

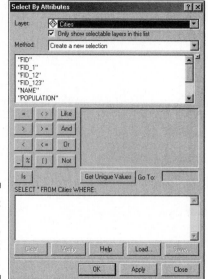

Figure 11-1:
The GUI
used in the
ESRI ArcGIS
product.

Knowing the size of each polygon

When you enter polygon information in a GIS, the software often calculates each polygon's area and perimeter, and then stores the calculations in associated database tables. Beyond those simple measures, you can find information about distances between polygons, orientation, roundness or other shape values, major and minor axis lengths, and so on.

Working with concentrations of point objects

With today's GIS software, you can create polygons based on the concentrations of point objects. For example, suppose you map individual crime occurrences and notice concentrations in particular neighborhoods. You can group these concentrations into separate polygons (by neighborhood) and use those polygons to compare crime data to neighborhood economic conditions. Some GIS analysts call this *hot spot analysis.*

Maps can contain many related polygons. Think about these related distributions:

✔ Zoning polygons coinciding with land values

✔ Vegetation patterns coinciding with animal ranges

> ✔ Average income by census district compared to education level
>
> ✔ Land-use patterns in 1997 compared to land-use patterns in 2007
>
> ✔ High cancer areas compared to hazardous waste sites

Reorganizing data

You can modify polygonal categories, values, or boundaries so that you can control the content of the maps you use for analysis. When you do so, you don't change the data; you just organize it differently. And you have various ways to reorganize map content, including

> ✔ **Clumping, or *aggregating*, categories:** For example, you might change categories such as Pine, Oak, Birch, and other tree types into a single category called Forest (see Figure 11-2). You might want to use this larger category to compare the position of the resulting polygon to areas of land owned by the federal government.

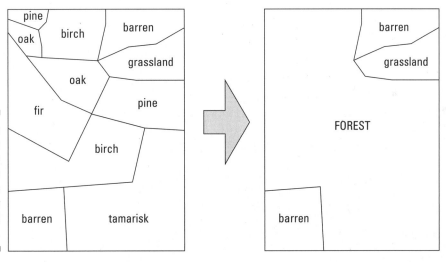

Figure 11-2: You can aggregate tree species data into a category called Forest.

> ✔ **Changing data measurement scales:** Suppose you have maps of *land cover* (a fancy name for what's on the land), and you want to change these nominal data (farmland, bare soil, grassland, forest, abandoned, and so on) to ranked data based on the degree to which the land cover types allow soil erosion. To make this change, you assign each category a number. In this case, you want the numbers to show the susceptibility to erosion. Because bare soil is probably the most susceptible, you give it a value of 1. Abandoned land gets a value of 2; farmland gets a 3; forest gets a 4; and grassland gets a 5.

Locating 2-D Map Objects

You can search for polygons in six ways: by (1) category, (2) data level, (3) size, (4) arrangement, (5) orientation, or (6) position. Each method accomplishes a specific goal. For example, you might want to search for a certain type of forest (category), or for forests that are grouped together (arrangements), or forests that occur in linear strips oriented north and south (orientation). The following sections offer examples that use each of the six methods.

Searching based on category

Most objects, whether stored at ordinal, scalar, interval, or ratio scales, typically have names associated with them. If you aren't interested in how much or how many, then you want to search by category name. A common type of nominal map is the land-cover or land-use map. Such maps depict many types of land cover or land use and show these features as categories in the legend (see Figure 11-3).

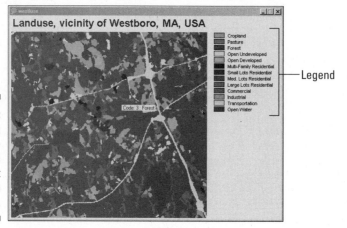

Figure 11-3: Selecting the categories of land use that you want to display.

Some systems enable you to access nominal categories directly from the map legend, and others give you a GUI based on SQL. (I talk a bit about search strategies in the section "Extracting specific information," earlier in this chapter.) The results of searching from the legend or with a GUI differ:

✔ When you search from a legend, you usually double-click the legend itself and pull up a list of options that you can pick from. Essentially, though, you're simply reclassifying or hiding categories (to remove them from view, see Figure 11-4), not really creating a new map.

✔ The GUI approach usually asks you to search categories by using a series of logical or mathematical operators. You use the GUI to find the categories you're interested in, and the search creates a whole new map for you.

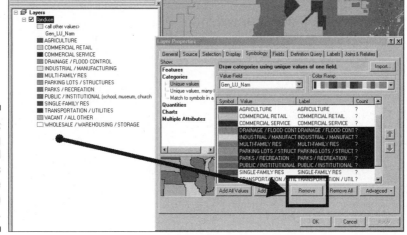

Figure 11-4:
Changing
the nominal
categories
of land use
displayed on
the map.

Get to know your GUI search tool as soon as possible — you'll use it often in your GIS work. Start with simple searches, and when you get comfortable with it, move to more complex searches. Have fun with it!

Finding polygons based on level

Many polygons come in *ordinal* (ranked) categories. The user might have created those categories for a particular use in which the scale is context-dependent *(scalar),* or another user might use more universally understood categories, such as predefined values for high, medium, and low housing densities.

In a land-use map like the one shown in Figure 11-3, you can select Residential as a single category based on nominal data. But suppose that you want to be more specific. The map data might offer the residential land uses not just as a single category, but also at different levels based on the number or type of housing units. Figure 11-5 shows the map legend with a breakdown of residential land use in levels of multi-family units.

You might want to buy a home in a low-density residential neighborhood (one that doesn't have many multi-family units). By selecting a polygon that represents low density levels from among all the other housing polygons, you can determine which parts of town would suit your living needs. Figure 11-6 shows how you can use a built-in GUI search function to select the levels of housing density from within the GIS database and create a new map from these more specific categories.

Searching for different attributes within a map legend doesn't create a new layer. Most often, searching for attributes allows you to quickly change the display, but that's about it. If you want to save and use the resulting map, you must use the GUI.

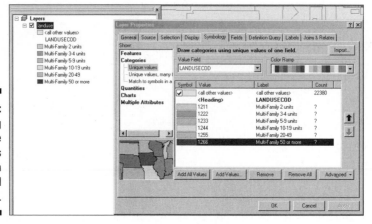

Figure 11-5:
Displaying land-use polygons based on ordinal categories.

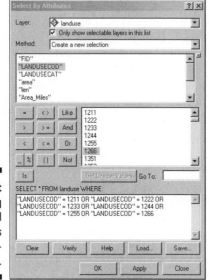

Figure 11-6:
Selecting ordinal categories using land-use codes.

Looking for polygons based on value

Suppose that, looking at a land-use map, you have a layer that shows the value of the land in dollars per acre. Such ratio-scale data are easy to separate from one another because they are numeric values. If you want to buy land that's priced at $100,000 or less per acre, for example, you need only select all land parcels whose attributes match the value less than or equal to $100,000 (see Figure 11-7).

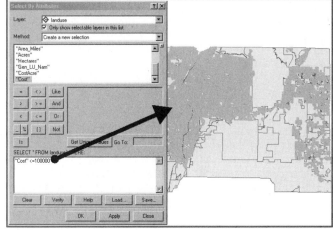

Figure 11-7: Selecting polygons based on value.

Maps can trick you. Before you search for polygons based on value, be sure that you're searching for the correct unit value. If you're looking for land based on price, for example, you're probably looking for price per unit area, not total price. If your map shows total price, you need to do some math to figure out the price per unit area.

Locating polygons based on size, shape, and orientation

Polygons come in different sizes and shapes. Asymmetrical polygons are aligned in a particular direction (north, south, east, west, northeast, southwest, or some other combination). You can find polygonal features based on their size, shape, and alignment, but some preparation must take place first. Don't worry — the GIS software does the prep work. You just have to find and use its tools to do your search.

If you're buying land, you need to know more than just how much it costs. You also need to know how big it is. After all, if you can get land for less than

$100,000 per acre but you need 5 acres, you must specify the amount of land (the size of the parcel) in your search. The good news is that both raster and vector GIS calculate the size of the polygons when you input those polygons into the database.

So, the software does the first part you need — determining the polygons' sizes — for you. You can go into the software and find the polygons based on some set criteria that you define, such as >5 Acre. In fact, you can combine the >5 Acre search with the <=$100,000 per Acre search to find just the set of lots that you might want to buy. The results of a combined search look something like the map in Figure 11-8.

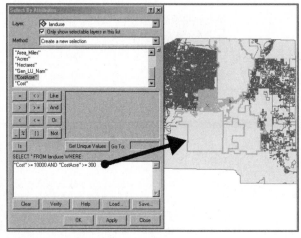

Figure 11-8: Locating polygons based on size and price per acre.

With GIS, you can use the many maps that you create to create even more maps, narrowing your search while you go. These steps show an example of how this narrowing-down process works:

1. **Use your GIS interface to apply a SQL query that selects properties within a particular price range.**

   The resulting display highlights the properties that match your search.

2. **In that same map (with the properties already highlighted), apply another query to select properties of a particular acreage.**

   When you get the results, you probably have more than one possible parcel to choose from. Some parcels might be much bigger than the 5 acres you need. You can modify your search to find only those parcels that are between 5 and 7 acres. (In fact, you could have done that in your first search.)

3. **Save the map you made (in this example, all the land parcels between 5 and 7 acres in size and priced at less than or equal to $100,000) to a new map.**

   But even then, you might have too many choices. Luckily, you can narrow your search even more. You notice that not all the parcels are the same shape: Some are long and skinny, and others are square and blocky. And you decide to select parcels on the basis of shape.

4. **Visually inspect the polygon shapes that represent your selections up to this point and choose from what you see.**

   Perhaps you decide that you like the blocky parcels, as opposed to the ones that are long and skinny, and you want to narrow the selected properties accordingly. The good news is that you can spot (within your ever-narrowing set of properties) what you need right away.

   You can calculate shape in GIS, but you don't really need that level of technical detail right now. In fact, most GIS software is only beginning to deal with shape analysis. Such power will no doubt be increasingly available as GIS software developers create new programs for shape analysis.

   Most GIS software allows you to draw boxes that touch the map features you want (which you can read about in Chapter 10), so you don't have to do any fancy calculations. Just point and click.

5. **Draw boxes that intersect the polygons you see that match your criteria.**

   Besides not wanting long, skinny parcels, you might want parcels that are oriented north and south, rather than east and west. Choose selected polygons that are the right shape and oriented just the way you want them.

When you're done, you can save this multi-step search as a map, as well. So, take the lead with your search criteria! Instead of relying solely on the GUI and SQL, you get to use your own powers of observation and select items based on what you see.

Finding polygons based on location and position

While you search for polygons, you may want to consider where they're located and how they're positioned. You may, for example, want your parcels within a certain distance of town so that you don't have to travel too far every day. And maybe you want to find land near a lake so that you can go fishing without having to drive for hours. You might also want your land to be on a hill rather than in a valley, or in the foothills rather than in the mountains.

You can perform searches based on location and position by using the GUI query tool. In Chapter 12, I show you how to measure distance. Most GIS software even has a tool you can use to select objects that are touching (adjacent), so you know exactly what parcels of land are next to each other.

Like you do when locating parcels with particular sizes, shapes, and orientations, you can easily estimate distances from one polygon to another and use the graphical query tool — the little box that you drag over the polygons — to select the ones you want.

Defining the Groups You Want to Find

Your project goals almost entirely control how you choose the groups of features you want to find. No formula exists for choosing groups, but you can ask yourself some basic questions. These questions are based on the types of data you have, the theme or themes you're interested in, and what feature properties and underlying geography you need to know about.

Looking for common properties

To find groups of geographic features — whether they're points, lines, polygons, or even surfaces — you first need to know what properties they share. You already do this type of grouping naturally when you decide on the types of movies you like to watch (categories); whether to order a small, medium, or large pizza (rankings); or how much money to spend for various types of entertainment (ranges). You can use the same thought process to select polygons by

- ✔ **Category:** The same type of object. You can select all oak trees in a map of trees, single-family dwellings in a map of housing, highways in a map of transportation, agricultural lands in a land-use map, or alluvial fans in a landform map.

- ✔ **Rank:** You can search for groups of features based on differences in rank. For example, separating cities by population ranges or selecting land units by high, medium, and low risk for earthquake or avalanche.

- ✔ **Scale:** If you have point, line, area, and surface data measured at interval or ratio scales, you can group those features pretty simply. You can select all stores (points) that have a particular dollar value of sales, roads (lines) that have selected specific traffic volumes, land parcels (polygons) that fall within a range of value per acre, or topographic surfaces that have slopes at a given range of elevation that change over a certain distance.

✔ **Spacing:** You might want to select objects based on what they're near, how far apart they are, their average spacing, or whether they're evenly spaced, dispersed, or clustered.

You can apply all these methods to properties of the intervening space (the space between features), to the features themselves, and to those features' shapes.

Looking for common positioning

Sometimes, you may find that geographic features seem to occur in some places and are totally absent or very poorly represented in other places. Such differences in density are not necessarily just statistical anomalies. Instead, the features tend to occur in certain places because of some underlying process or control that causes them to occur in those places and not in other places (that is, the distributions are determined by another distribution or process). In other cases, the relative density of the features changes in an observable, statistically predictable fashion.

Here's a quick list of some features that might exhibit patterns related to some form of controls or outside forces:

✔ Earthquakes and volcanoes (related to the locations of continental plates)

✔ Vegetation types (related to soil types, slope and aspect, or herbicides)

✔ Mobile homes (related to zoning)

✔ Vandalism (related to neighborhood type and/or gang activity)

✔ Winter tourists (related to winter temperatures)

Because these features occur in one area versus another area, you have to deal with a new problem — that of determining why such differences exist. To a geographer, such patterns form questions. These questions might include:

✔ **Is the pattern real or a product of sampling (or lack of sampling)?**
Some striking distribution changes aren't necessarily accurate: They may be a result of lack of sampling. First, check whether your distribution reflects sampling methods, mistakes, omissions, or even oversampling in one area versus another. You need to know your data (a good reason to produce your own database when you can) or establish a way to spot-check your data for accuracy. Check out Chapter 8 for more on sampling.

Don't just assume the patterns that occur on maps are accurate. Check the metadata for information on methods and sampling procedures, as well as any other information that allows you to verify the data's reliability.

✔ **Is there a distinct difference between places where features occur?**
Sudden changes most often indicate an equally abrupt change in some
other factor that either controls or contributes to the existence of the
features. And if your distribution changes gradually, a process is probably controlling that change. The process may occur naturally. Suppose
you notice that the cotton plants that have grown in a certain place for
a long time are starting to die off. The growing process takes nutrients
from the soil, and the cotton plants — growing in one location year after
year — have depleted the nutrients in that area.

If the distribution changes gradually and in a particular direction, this
change may reflect the direction of change in an underlying process. For
example, density may seem to change from the middle outward or from
one direction to another. A change from the middle outward can indicate dispersion from high concentrations to low concentrations, while
a directional movement can indicate movement toward more favorable
environments, sources of nutrition, and so on. For example, a herd of
bison may migrate toward new sources of plants for grazing.

Abrupt changes in distribution usually indicate the presence or absence
of a controlling factor.

✔ **What area of the map does a set of points or lines occupy?** When
distributions are spottier and change from dense to sparse in a more
variable way, you may find it difficult to figure out whether the patchy
distributions (each of which might be either clearly defined or gradually
changing) are related to one another functionally. But you absolutely
must make this determination because your conclusion based on a distribution often assumes that the distribution maps are correct. When
you create a grouping, having experience observing how distributions
naturally cluster helps you determine where to draw the lines around
groups of features. Some specialized software packages use sophisticated statistical techniques to make these grouping decisions, but most
general GIS software is only just catching up.

One basic method of enclosing distributions into a single polygon is
almost like using shrink-wrap. You decide which points you want to
include, and the software draws a polygon by connecting the outside
points, creating the *least convex polygon* (like the one shown in Figure
11-9). This simple and easy method groups features rather nicely for
most distributions. If you need more accurate methods, you can search
the spatial statistics packages included in modern GIS software.

✔ **Does the area of distribution correspond to any other feature or distribution?** When you begin to look at distributions and see patterns in
those distributions, you naturally start asking whether one pattern is
related to another, whether one causes the other to exist, and to what
degree these patterns correspond. Chapter 15 addresses this question
in detail.

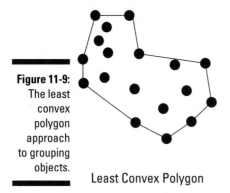

Figure 11-9:
The least convex polygon approach to grouping objects.

Least Convex Polygon

Grouping by what you already know

Even the seemingly mundane task of finding the location of geographic features lets you create new maps that answer questions and allow you to make powerful decisions. When you engage your GIS to find and group objects, you base your search strategy on your project goals (what you need to know), but also rely on what you already know to help direct your search.

Suppose that you're an *epidemiologist* (a scientist who studies infections). You need to understand the distribution and spread of West Nile virus. West Nile virus is spread by mosquitoes, which need standing water to breed. You must therefore identify places where outbreaks of the disease occur most frequently so that you can identify places where you can hopefully eliminate the cause (the infected mosquitoes).

This basic scenario gives you all the building blocks you need to formulate your questions and decide what types of data groups you're looking for. Just follow these steps:

1. **Identify your target objects and a search strategy to find them.**

   To satisfy the goal in this example (eliminating infected mosquitoes), I want to locate and group two things: patients with West Nile virus and mosquito breeding grounds. You select, qualify, and group your data basing your strategy on what you know:

   • **To find patients:** Look for point data in the form of the locations of patients. You need to group these patients into two classes: those who have contracted West Nile and those who haven't.

   Further qualify your patient data by selecting out the class data that occur in high density — for example, more than two cases of the disease per square mile. Now, you have a set of points that illustrates high versus low density of West Nile virus throughout your study areas. By narrowing down the patient data groups, you may begin to see patterns emerging that suggest selected areas that might have high numbers of mosquitoes.

- **To find mosquito breeding grounds:** It's really hard to map mosquitoes as point objects (they're pretty little, and those GPS units make flying awfully difficult). But you do know that mosquitoes require standing water to breed. Standing water can be anything from ponds to puddles, pools to tires with water in them, and you just happen to have data on all these features. You need to group the polygons that are likely to have standing water and those that aren't because you assume that areas with lots of standing water (a higher density of standing-water polygons) are much more likely to be mosquito breeding grounds.

  So, you regroup your polygons based on how clustered they are, in relation to one another, using the feature's geography — defined here as the distribution of the polygons — to give you a ranking of high, medium, and low potential for breeding mosquitoes.

2. **Create maps showing your newly grouped target objects.**

   In this example, you have one map that shows concentrations of West Nile virus cases and another map that shows different levels of mosquito breeding potential.

3. **Overlay and compare the maps showing the grouped objects.**

   If your assumptions are correct, you see higher concentrations of West Nile virus near the locations having a higher potential for mosquito breeding grounds. This correlation suggests exactly where you can begin a mosquito abatement program to reduce the risk of infection.

   But you might also see situations in which higher outbreaks seem to occur where mosquito-breeding potential is medium or even low, suggesting that you need to do a further examination of geographic features that might account for these inconsistencies.

This example shows you how to perform a simple analysis of the types of objects (in this case, points and polygons), the measurement levels (nominal), and their geographic distributional attributes (concentrations of infections and of standing water sources).

You can use the same approach, basing your grouping strategy on what you need to know plus what you already know, to satisfy many other kinds of GIS projects. In any case, many features you want to group share both functional and spatial relationships. The functional relationships (such as cause and effect) help you decide which features you want to search, and the spatial relationships allow you to identify the degree of *spatial co-occurrence* (where different geographic features seem to occur in the same locations). Chapter 16 talks about how you can use GIS software to make these comparisons without guesswork.

Part IV
Analyzing Geographic Patterns

The 5th Wave   By Rich Tennant

"Using GIS software, I was able to overlay topographical maps of Afghanistan, Kuwait, and Montana on top of each other. Now, if you'll notice, it kind of looks like a winged pony. See the tail...?"

In this part . . .

The most advanced functions of GIS lie in the systems' capacity to measure geographic features, including distances, areas, networks, surfaces, and all their related characteristics. In this part, you find out how GIS performs individual tasks, such as finding shortest or fastest routes, analyzing visibility, assessing fluid accumulation and flood zones, and many others. Also, you discover how to combine these individual operations to create complex, real-world models, which are (in my opinion) the most rewarding pinnacle of the GIS analyst's art and science.

Chapter 12

Measuring Distance

. . . .

In This Chapter

▶ Measuring absolute distance

▶ Calculating relative measurements

▶ Figuring out distance based on function

. . . .

*Y*ou can measure anything you put on a map. You use map measurements to figure out how far you are from one place or another, how much land you own, how much ore you might expect to get from a new mine, and many other measurements. You can measure heights, widths, depths, lengths, areas, and volumes. You can also compare measurements as ratios — for example, ratios of length to width, perimeter to area, or height to distance.

Measuring geographic features is one of a GIS's strengths, and it does the measuring very quickly! Because you can measure many features quickly, you can use those measurements to help you make decisions about what route to take (by comparing distances between features), whether one place is better than another for mining (by calculating the volume of a specific feature), or where you're likely to find the best ski slopes (by evaluating a feature's height against its length).

In this chapter, I show how GIS gives you the power to measure all sorts of features. First, I explain the measurements that GIS software generates by default (such as grid cell size), and then I describe how the GIS uses these numbers to calculate other measurements (such as distance between objects). I also tell you about the difference between relative and absolute measurements, and when you might want to use each. I explain how you can use measurement tools to describe how features are distributed relative to each other and to other surrounding features. And finally, I describe how to apply distance to make entirely new polygons based on the distances you measure.

Taking Absolute Measurement

When you enter data into a GIS, the software keeps records of those points, lines, and polygons. More important, it keeps a record of the X and Y locations of your vector or raster objects. It stores the size of the grid cells and the lengths of the lines. Having these records saves a lot of time when you calculate distances.

The measurement of absolute distances, whether between objects or along lines, is computed differently for raster and vector data structures. The following sections describe these measurement methods and how the computer implements them for both data types.

Finding the shortest straight-line path

Finding the shortest straight-line path between any two points is the simplest method to measure distance. Often called "as the crow flies" distance, this *Euclidean method* makes two assumptions:

1. No obstructions block the path between the two points.

2. You don't need to travel along established paths and roads.

The following sections describe how the GIS software enables you to find the shortest straight-line path between two points that exist under a variety of circumstances.

Finding the shortest distance on a flat surface

If you're working on a flat surface, the shortest distance between any two points (a straight line) is based on a well-known mathematical equation ($c^2 = a^2 + b^2$), the Pythagorean Theorem (named after the ancient Greek mathematician Pythagoras). The Pythagorean Theorem is based on the use of a geometric form called a *right triangle* — any triangle that has one corner that makes a right-angle turn (90 degrees). The longest side of this triangle, the one opposite the right-angle bend, is called the *hypotenuse.* Figure 12-1 shows an example of a right triangle.

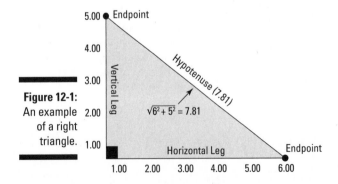

Figure 12-1:
An example of a right triangle.

Remembering sines and cosines

When you work with planar coordinates for measuring distance, you need to know the X and Y values for each point. So, you can convert spherical coordinates to planar coordinates as follows:

✔ **The X location:** X = r × cos(a) × cos(b)

✔ **The Y location:** Y = r × cos(a) × sin(b)

Here are what the letters stand for in the preceding transformations:

✔ **r: The radius of the Earth**

✔ **a: The latitude**

✔ **b: The longitude**

Seeing sines (abbreviated sin) and cosines (abbreviated cos) may make your eyes water, but you can remember what these terms mean by taking the pneumonic device SOHCAHTOA and breaking it into three parts — SOH, CAH, and TOA:

✔ **SOH:** Sine equals Opposite over Hypotenuse

✔ **CAH:** Cosine equals Adjacent over Hypotenuse

✔ **TOA:** Tangent equals Opposite over Adjacent

So, to calculate sines and cosines, you really just have to divide. After you transform all your spherical coordinates to planar coordinates, you can apply the formulas that you use to calculate straight-line distance to your new X and Y coordinates. But I have even better news — the GIS does all this for you!

Each of the two points (that you want to measure the distance between) is defined by an X coordinate and a Y coordinate along number lines. To locate each point, you have to move horizontally and vertically along the number lines from the origin point (where the right angle forms) to the values that define your points. These movements along the number lines define the straight legs of your right triangle. You can think of the distance between those two points as the hypotenuse of your right triangle.

When measuring real Earth distances in a GIS, the software calculates the hypotenuse distance using the X and Y values of two points in a modification of the Pythagorean Theorem called the *distance formula.* You'll encounter this distance formula many times in your GIS work.

When you need a quick online refresher on the distance formula, visit www.purplemath.com/modules/distform.htm.

Finding the shortest distance on the spherical Earth

The GIS can also calculate distance between a set of latitude and longitude coordinates on a spherical Earth. You may need to find the shortest distance between two points on the globe. This calculation is called a *great circle distance* and uses a geographic grid. The calculation of latitude and longitude distances has some very strong similarities to calculating distance by using flat (planar) coordinates and the distance equation. (In Chapter 2, I describe a bit about converting a spherical Earth to its representation on a flat piece of paper.) This process, called *map projection,* involves mathematics because it converts spherical coordinates to planar coordinates.

Conversely, no space on the Earth is totally flat, so calculating the distance between any two points requires some determination of the spherical coordinates. Your GIS is already set up to perform these calculations for you. Using trigonometric formulas and the X and Y (planar) coordinates of geographic features, your software can easily change back and forth from spherical geometry to planar geometry. If you really want a peek at the math behind this conversion, check out the sidebar "Remembering sines and cosines."

Measuring distances in grid cells

Measuring distances in raster systems (using grid cells) is a bit different than measuring distance in vector systems, both conceptually and computationally. Many raster GIS packages allow you to calculate measurements in both flat coordinate systems and the geographic grid (latitude and longitude coordinates). The alternative to measuring by coordinates is counting up grid cells.

If all the grid cells in the path you're measuring are edge-to-edge *(orthogonal),* you can easily figure the distance the path covers. Just count up the grid cells and multiply that number by the width of the grid cell. For example, if you count 150 grid cells from one point to another and each grid cell is 20 meters across, the total distance is 20×150, or 300 meters.

This very simple approach gives you pretty accurate results, but only if all the grid cells are orthogonal or all distances are exactly lined up by the cardinal directions. In real distances, you seldom get such an exact configuration. If you're working with a raster system in which you want to measure distances that are diagonal, you need to measure grid cells that are diagonal to one another, at least for some portion of the distance (as shown in Figure 12-2).

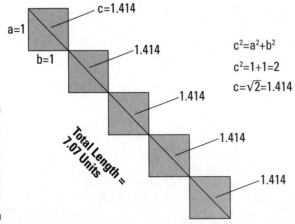

Figure 12-2: Measuring diagonal distance in grid systems.

c=1.414
a=1
b=1
1.414
1.414
1.414
1.414
1.414

$$c^2=a^2+b^2$$
$$c^2=1+1=2$$
$$c=\sqrt{2}=1.414$$

Total Length = 7.07 Units

Measuring Manhattan distance

When I was a young lad in Minnesota, I used to like to go through large department stores on my way home to get out of the stinging cold and, ideally, shorten my path. But actually, the path I took was just as jagged as if I'd stayed outside. Although I did manage to warm up for a few minutes, I probably didn't make my trip any shorter.

Imagine you're in the downtown business portion of any large city, such as New York City's Manhattan, where the streets are laid out like a grid. Each block is the same size, and tall buildings prevent you from moving anywhere but along these blocks — unless you can leap tall buildings in a single bound, of course. Likewise, cars must follow the streets that separate these blocks. The distance from point A to point B, unless both points are on the same block, is going to be longer than if you could travel "as the crow flies," as I discuss in the section "Finding the shortest straight-line path," earlier in this chapter. Because Manhattan is an excellent example of this restricted movement, geographers refer to distance measurements in this environment as *Manhattan distance.*

Figure 12-3 shows a hypothetical city with the same type of structure as downtown Manhattan. To travel from point A to point B, you need to follow the edges of the blocks. By moving along the blocks, you have to travel a much larger distance than if you could travel the straight-line distance. This grid structure allows many possible paths from one point to another, but each path covers essentially the same distance.

Figure 12-3:
Measuring Manhattan distance moves from point A to point B along an unobstructed path.

To calculate Manhattan distance, you use a modification of the distance equation, but conceptually, the calculation is really much simpler than that. Because each of the edges for each block is a straight line segment, you need to know only where the X and Y coordinates are for the ends of each line segment. Most GIS software automatically stores that information when you first input the data. You can easily calculate each line segment's length using the distance equation included in the GIS software. To determine the total Manhattan distance, you simply let the software add together each individual line segment's length.

Calculating distance along networks

You encounter many types of routes in GIS work — including footpaths, animal tracks, rail lines, intercity roads, and street patterns — that look nothing like Manhattan. Each of these networks typically indicates an imposed structure on distance traveled. To calculate distance along these networks, the GIS measures vector segment lengths or grid-cell lengths for vector and raster, respectively.

- ✔ **Vector:** The GIS first retrieves the X and Y coordinates of each line segment's endpoints (stored at input) and then uses the distance equation to determine each line segment's length. Finally, the GIS adds together all the individual line segment lengths.

- ✔ **Raster:** The GIS finds each orthogonal grid cell's width and then adds those to get the total length. When grid cells are diagonal to each other, the lengths are measured as the width multiplied by 1.414 (the diagonal distance of a grid cell), as illustrated in Figure 12-2.

You can measure linear features in raster, but you can both represent and measure features more accurately in vector.

In Manhattan distance, you don't gain anything by choosing a different route. With most irregular networks, however, you can choose many routes from point A to point B, and these routes often have wildly different distances. Your software can measure all these possible routes and find the one that has the shortest distance by using the shortest path algorithm (an *algorithm* is like a recipe to solve a problem). Chapter 16 covers the shortest path algorithm in more detail.

Working with buffers

Nearly all GIS software has a set of special distance functions designed to selectively calculate distances around existing geographic features. These operations are called *buffering* (no, not like the aspirin), and the result is called a *buffer*. The software measures outward a certain distance from a point, line, or polygon (even the base of a topographic feature), and it then converts that whole area to a polygon.

A lot of real-world features contain buffers — prescribed, natural, or enforced distances that surround them:

✔ Easements around power lines, railroads, and so on

✔ Road frontages and street setbacks

✔ River vegetation, beaches, and drainage-way vegetation

✔ Police lines, safety zones, and concert barricades

✔ Earthquake zones and 100-year flood zones

The user bases each corridor on rules, natural occurrences, safety, utility, security, and any number of other factors. The GIS allows you to implement these conditions and rules in a number of different ways, as described in the following sections.

Placing buffers around points

You can place buffers around point objects, such as wellheads. You can decide to make this type of buffer either a single distance measure or in multiples (as shown in Figure 12-4). So, if you put a buffer around a wellhead as a safety corridor, you can actually create several levels of safety corridor — 100, 200, and 300 meters, for example. These might be classified as modest, medium, and high risk based on a safety officer's experience.

Placing buffers around lines and polygons

Buffers around lines and polygons give you three options that point buffers don't:

✔ **Variable buffer:** By using the GIS *variable buffer* function, you can select the size of the barrier for each portion of a line or polygon feature, instead of being forced to use the same-sized buffer for the entire feature. For example, in a river network, the trunk stream probably has more vegetation on it than its tributaries do, so it has a larger corridor than its tributaries' corridors (see Figure 12-5).

✔ **Setback buffer:** You can determine whether a buffer will be the same size on either side of a feature. In some cases, this difference is extreme — the buffer exists on only one side of a feature. This type of buffer is called a *setback buffer* because it's "set back" in only one direction. You can position a setback in either direction from a line feature (for example, the center of a street). For polygon features, the setback buffer is usually established (set back) from the perimeter to some distance inside the feature.

✓ **Bidirectional buffer:** Most buffers measure distance outward. Setback buffers can measure distance inward. One advantage of doing a buffer around an area feature is that you can have a buffer going both directions at once. Most GIS software allows you to select this option so if you want to measure a buffer distance both inward and outward from the outside perimeter of an area feature, you can.

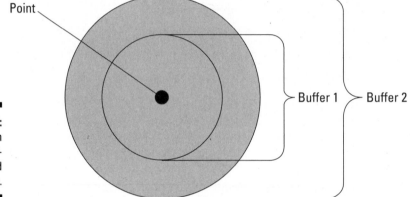

Figure 12-4:
You can create buffers around points.

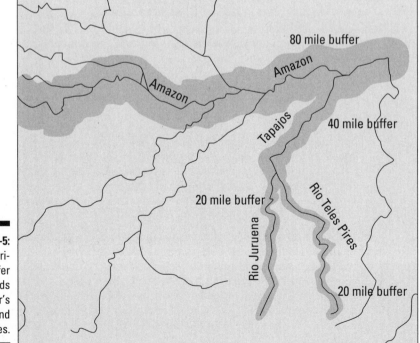

Figure 12-5:
This variable buffer surrounds a river's trunk and tributaries.

Establishing Relative Measurement

Sometimes, you don't need to know the exact distance from a reference point to a feature; you just need to know whether that feature is close to or far from the reference point. In some cases, a feature might actually be touching the reference point, a position that's called *adjacency*.

Relative measurement is the type of information you need every day. If you ask about the distance to something, you don't always expect, nor even want, a precise answer. For example, say you're in a new town and stop to ask directions — "How far am I from the courthouse?" You receive the answer, "It's about 766 meters east south east as the crow flies." This answer gives you an idea of where it is and how far, but you were really looking for something like, "It's just a short drive down this street and around the corner." In other words, you just wanted to know whether you were close or whether you had to drive for quite some time.

You can use relative measures in many ways:

✔ Selecting among different locations

✔ Deciding whether you have time to get someplace

✔ Knowing whether features such as towns are going to interact with each other

✔ Determining whether you can find sufficient beachfront property for a resort

Adjacency and nearness

When your reference point is so close to a feature that the distance between them is essentially zero, you call that distance *adjacency*. The distance measure for adjacency involves how many features are at a distance of zero from your reference point.

Because relative measurement requires a starting point, you can easily determine whether something is near or far by creating distance buffers based on your definition of near or far. Current GIS technology can easily measure distance, but it can't make decisions about what you consider near or far for a given application. Suppose that you want to know which parcels of land are near (within a given specified distance of) a lake. You follow these basic steps to define *near* as within a quarter mile of the lake:

1. **Create a buffer that's a quarter of a mile from the polygon you select.**

   In this example (see Figure 12-6), I select the Lake polygon using the Select tool and use the Buffer tool to measure a buffer that's a quarter of a mile from the outside of the polygon.

2. **Use the Adjacency tool to select polygons that touch the buffer.**

   The software selects which polygons come in contact with any part of the distance buffer polygon you created (see Figure 12-6).

You can use a similar method to determine which polygons are actually touching, or adjacent. In this case, the GIS software doesn't need to create buffers to measure anything. Instead, it uses that powerful data structure — a topological data model — with the idea that the data inside the computer already have the necessary information to decide which polygons are connected to which others. (See Chapter 5 for more about the characteristics of topological data models.) You just need to issue a command telling the GIS which polygon is your target polygon and then ask it to find all the polygons that share lines with that polygon (refer to Figure 12-6).

Figure 12-6:
Polygons
that touch
a distance
buffer.

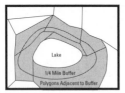

Separation and isolation

When you measure separation and isolation, you compare the absolute coordinates of one geographic feature to several other features so that you can determine the average distance from a feature to other features. This form of distance measure is especially useful in fields such as ecology, in which scientists need to determine relationships such as whether certain patches of vegetation are close enough for species to interact.

To calculate isolation, the software finds all the coordinates of the target polygon (a vegetation patch, for example) and measures the distance to the nearest points for all the other polygons in the area. Another common way of determining isolation is to calculate the average distance between the target polygon's centroid and the centroids of all other polygons.

No matter which method your software uses, it essentially calculates a measure of average distance among polygons.

Containment and surroundedness

You generally think of measuring distance from one thing to another thing. When your reference point is inside a polygon, its distance from the edge of that polygon is measured inward toward the reference point rather than outward toward other features. This idea is called *containment.* Containment can help you figure out what features are inside other features.

To measure containment, the GIS essentially measures a distance inward from the edge vertices (the corners) of a polygon or area feature, determining whether the X and Y coordinates of your object are all contained within the limits of the X and Y coordinates of the polygon that you examine (see Figure 12-7).

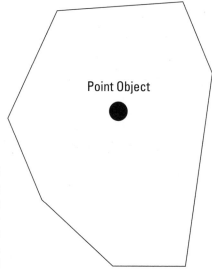

Point Object

Figure 12-7:
A point object contained in a polygon.

Knowing whether an object is contained in a polygon can help you answer these kinds of questions:

- ✔ Does the property I'm interested in fall within a certain school district?
- ✔ Does a particular wildlife species live in a certain habitat type?
- ✔ Do I live inside a 100-year flood zone?
- ✔ Which voting district do I live in?

To answer questions like these (for example, about voting district), you usually let the program compare the data from two different maps (for example, a street map and a precinct map) because both kinds of data (street addresses and precinct boundaries) don't appear on the same map. Still, to determine whether the distance would be inward (contained) or outward (not contained), you use the same calculations that you do for a single map analysis because the software essentially compares the X and Y coordinates of the points on one map (the street addresses) to the polygon coordinates (the precinct boundaries) on the other map.

You can assume that a point object exists either inside or outside a given polygon because it has zero dimensions. If you ask whether a line feature (such as a highway) is in a polygon (such as your state) or another polygonal feature (such as a park) is in a polygon (such as your city), you add dimensionality that complicates the answer. You can have a situation in which part of a line or polygon is within another polygon. GIS can help you determine whether these objects are only partly in a given polygon by simply comparing the locations where the two sets of features intersect.

Measuring Functional Distance

Functional distance is a measure of space based on how that space is used, rather than its absolute measurement. Functional distance recognizes that you don't always want to know exactly how near or how far, but rather what traveling that distance might cost you in time, money, calories, or even just emotional distress. Simply put, you use a different measuring device than for absolute measurement. Say that you ask the question, "How far is it to the interstate highway from here?" You might get a functional-distance answer of, "Oh, not far! Maybe a couple of minutes."

You can base functional distance on these kinds of factors:

- Expenditure of fuel, based on how long it takes to travel a distance
- How much a trip costs
- How much wear and tear you put on your vehicle
- How much stress the trip puts on the travelers

These factors can influence functional distance measurements:

- The ruggedness of the terrain
- Whether the roads are well maintained
- The amount and direction of wind

You can find as many functions of geographic space as you can find potential users. The following sections describe various forms of functional distance.

Anisotropy (whew!) — non-uniformity

GIS sometimes gives you a chance to impress your friends because you know all these cool-sounding words that they've never heard. One such word is *anisotropy,* which is the opposite of isotropy. *Isotropy* is the property of a totally smooth surface that has no paths, bumps, wrinkles, vegetation, or obstructions of any sort. If you find an isotropic surface, let me know because it's really just a theoretical property. Isotropy doesn't exist in nature, but it does give you a standard against which you can compare all your anisotropic surfaces.

A perfectly isotropic surface that has no obstructions and no slopes, and a totally smooth surface produces a situation in which you can travel outward from its center in any direction and end up at the same distance from the center as you would if you'd gone in any other direction. It's sort of like a series of concentric circles in which the circles show distance traveled at selected times.

When you want to measure distances from a point or area to all other locations throughout a surface, knowing that the surface isn't uniform (meaning not isotropic), the idea of anisotropic surfaces comes in. The roughness of the surface is a measure of its departure from the perfect isotropic surface. The rougher the surface, the more likely it is to produce *friction* (resistance to movement), as shown in Figure 12-8. The steepness, type of surface, or stability of the subsurface can cause varying degrees of friction. Factors such as the time of day, direction of travel (uphill or downhill), and how long the traveler has been traveling also influence non-uniformity.

Accounting for physical parameters

The most common factors associated with impeding travel are physical:

- ✔ **Slope:** The steeper the uphill slope, the more friction you experience. Think about how driving uphill into the mountains both puts a strain on your engine and draws heavily on the old gas tank. This same friction impedes bicyclists, hikers, and even wildlife.

- ✔ **Type of surface:** You can much more easily traverse smooth concrete than soft sand; you can walk along a well-trodden hard path more easily than saturated ground or mud.

- ✔ **Surface features:** The surface itself has the ability to support features that contribute to the impedance or friction value of the surface — such as vegetation. Think how much more easily you can travel along a soft grassy surface than along one that has thickets or trees.

Based on intangibles

Many *intangibles* — subtle, non-physical things — can contribute to differences in travel patterns and preferences. Pleasant scenery can contribute to a lower level of anxiety and may be a reason for taking a physically longer route than necessary. Many scenic routes slow travel from one place to another but add interest to a trip. State of mind can also help define perceived distance.

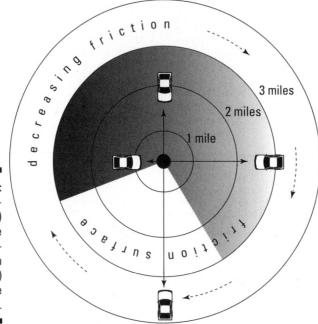

Figure 12-8:
Rough (high-friction) surface travel versus smooth (low-friction) surface travel.

Creating the functional surface

Whether you're trying to reproduce measurable physical parameters or the intangibles of functional distance (see the preceding sections), you may have difficulty assigning friction values — not because GIS is an immature technology, but because geographers might not know enough about the geography of friction surfaces to meet your needs.

Researchers have established a few measurable friction surfaces as standards for some physical features under selected situations. One of these, the Universal Soil Loss Equation, models the loss of soil during rain events. Within that equation, you need to assign a value to a surface type based on the average soil erosion factor for each soil. You then plug these generalized numbers into the equation and get a calculated estimate of how much soil will erode from a surface based on the surface type.

For GIS, the process of modeling friction surfaces is much the same as using the Universal Soil Loss Equation. (In fact, soil scientists and soil geographers have used the Universal Soil Loss Equation in GIS modeling.) The GIS modeler must encode any impedance to movement into a special map layer, often called a *friction layer*. The greater the friction value, whatever it's based on, the higher the value stored in the friction map.

In theory, you can quantify friction easily: `big friction = big numbers`, `little friction = little numbers`, `no friction = zero`. But, in reality, the process is a bit more complex than that. Ask yourself how you might give friction values to some of the following:

- Difficulty of crossing a stream

- Resistance of thickets for hiking

- Resistance of forests to deer movement

- The impact of fatigue on walking uphill

- The expenditure in calories of a cougar traversing a rugged mountain

- The impact of temperature on the movement of disease

And beyond these physical factors, imagine trying to model the impact of scenery on the apparent time of travel. How do you put in numbers friction values that have no standards and no measurements? You can use a scalar level of geographic measurement, in which you establish the scale of measurement. If you use a scale of 0 to 10, you probably know at least two values of your scale — 0 and 10. The 0 value indicates a condition that includes no resistance to travel at all, and 10 indicates that no movement can take place.

You need to decide where, within that range, other friction values might fall — by using experience, expert opinion, some measurements, surveys, or any number of different ways to gather some basic information. Unfortunately, you're stuck with a combination of guesses, experience, opinion, and luck. But although little may be known about friction surfaces, you can usually figure out the maximum (a barrier) and minimum (frictionless) values that allow you to place the upper and lower limits on your scale.

Calculating the functional distance

No matter how you define your numbers, your friction map must interact with another map layer that you use to find distances to and from places. Raster data models always work best for this kind of layer interaction. Figure 12-9 shows a raster data model. When you move from cell to cell in this data model without accounting for friction, you simply add the distance of the grid cells (1 width for orthogonal and 1.414 for diagonal, see the earlier section "Measuring distances in grid cells").

After you take friction into account, the computer must add not just the distance, but also the friction value before it can actually continue to calculate distance for the next cell. So, the higher the friction number, the more the computer has to add to the cumulative distance value before it graphically moves to the next grid cell.

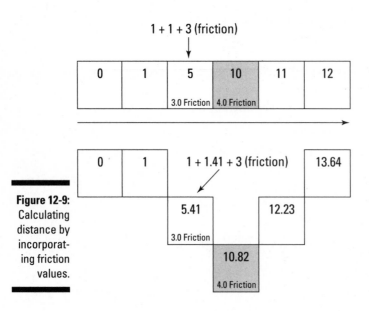

Figure 12-9:
Calculating distance by incorporating friction values.

Chapter 13

Working with Statistical Surfaces

*W*hen most folks hear the term *surface,* they think of the physical surface of the ground beneath their feet. Paper maps that show topography use elevation lines, and some globes actually have raised surfaces to indicate mountain ranges. With GIS, surface means much more than the physical topography. You can map and analyze many different statistical surfaces, which include features such as population density, crime rates, and cost of living. This chapter shows you the ins and outs of statistical surfaces, beginning with an overview of just what makes them tick.

Examining the Character of Statistical Surfaces

In GIS, the definition of *surface* is much broader than simply the physical surface of the land, or *topographic surface*. Surfaces can relate to other factors of the physical environment that help to create non-physical, or *statistical,* factors.

Non-topographic statistical surfaces come in two general forms — those relating to the physical environment and those relating to the human environment. This breakdown doesn't have any intrinsic value; it just provides a simple way to categorize non-topographic statistical surfaces. In the following sections, I refer to only those human environmental factors that relate to social and economic activities because these factors are easier to understand than the less tangible surfaces (such as those related to beliefs or attitudes, for example, whether people think economic conditions are improving) that occasionally appear in more obscure applications.

Here are examples of the two general forms of statistical surfaces:

- ✔ **Physical environment:** The most obvious physical statistical surface is topography, but many others are nearly as common and quite useful. Many physical maps relate to subsurface phenomena such as depth to bedrock, depth to groundwater, thickness of rock formations, depth to the ocean floor, and many others. Other physical surfaces are more closely related to climate and weather phenomena, such as temperature, barometric pressure, and precipitation.

- ✔ **Human environment:** Think about human environment in terms of socio-economic data. These data deal with people — their numbers, ages, races, education levels, and incomes — and with factors that affect people, such as cost, home values, and taxes. Socio-economic data can even represent disease, crime, attitudes, and belief systems. Because socio-economic data relate to geographic features that don't normally occur everywhere (that aren't continuous), analyzing such data often requires that you suspend disbelief (pretend that what really doesn't occur everywhere actually does) and treat them as proper, formal statistical surfaces.

All these factors and thousands more involve the measurement and mapping of values, each of which has a pair of X and Y coordinates (for location) and a third value — the Z value — which describes the surface. You can turn the values for each physical or human environment factor into a map of the surface that shows the highs and lows in Z value in different places (as illustrated in Figure 13-1).

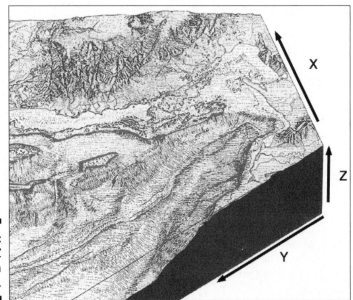

Figure 13-1:
X, Y, and Z values of a surface.

Understanding discrete and continuous surfaces

Surfaces can be discrete or continuous. A *discrete* surface results from mapping Z values that occur only in certain places in space. They don't occur at every point in the geographic space covered by the map. People, plants, animals, houses, and parked cars are all examples of features that you can map as Z values and that occur in discrete locations. Because discrete surfaces occupy separate and distinct places in space, they appear blocky rather than smooth, as illustrated in Figure 13-2.

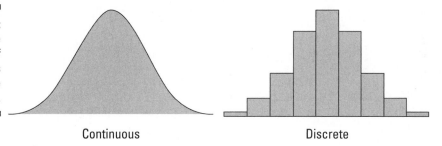

Figure 13-2: A side view of continuous and discrete surfaces.

Continuous   Discrete

Continuous surfaces (also depicted in Figure 13-2) include all features that occur everywhere — meaning that every part of the surface has a measurable value. Temperature, barometric pressure, and elevation all occur everywhere. Because they occur everywhere — at all points of the mapped area — an infinite number of them exist.

When you work with continuous surfaces, you work with an infinite number of data points. If you tried to include all this data in one GIS database, you'd need that database to have infinite storage capability — which just isn't possible. So, you sample from the total and hope that the sample represents the total population with an acceptable degree of accuracy.

The section "Sampling statistical surfaces," later in this chapter, covers the specifics of sampling procedures.

Exploring rugged and smooth surfaces

Surfaces can change slowly or abruptly, and you can classify these surfaces as rugged (as in rugged terrain) or smooth. The ruggedness or smoothness of a surface hints at characteristics of that surface. You can travel over gradual changes in topography much more easily than you can travel across abrupt changes, which signify rugged terrain; gradual changes in barometric pressure produce less variable weather patterns than abrupt changes; and gradual changes in soil nutrient levels result in less variable crop yields than abrupt shifts in soil nutrients. Figure 13-3 shows an example of rugged versus smooth terrain.

Figure 13-3:
Rugged (left) versus smooth (right) terrain.

Climbing steep surfaces

Like ruggedness, the *steepness* (change in Z value over distance) of a surface has an impact on surface characteristics. Steep slopes require more fuel consumption for uphill travel, steep changes in depth of an ore body change the amount of surface material that you must remove to access the ore, and steep changes in barometric pressure result in high winds. Figure 13-4 shows steep versus gentle topographic slopes.

Figure 13-4:
Steep (left) versus gentle (right) topographic slopes.

Determining slope and orientation

Slope is basically the measure of steepness for any statistical surface. It is essentially a comparison of how rapidly the Z values change as you move across the X and Y (horizontal) part of the surface. Any slope is positioned at a particular angle to the horizontal (the *gradient*), and you can measure that angle, figure out the nature or extent of the forces that formed it, and see how it affects other surface factors by its nature.

The concepts of slope and gradient apply most often to topographic surfaces, but not exclusively. For example, the slope of a population distribution indicates how quickly population changes from one place to another. Barometric-pressure slope often determines wind speeds. Rapid changes (slopes) in crime values typically mean that a local area has a location that experiences high levels of crime (a *hot spot*). You could come up with a quite large list of possible slopes and their explanations and/or consequences.

TECHNICAL STUFF

The math behind slope and orientation

Slope and orientation are complementary functions, and you generally calculate them simultaneously. Here's a brief look at how you do the math to determine slope and orientation:

✔ **Slope:** The amount of slope is based on the change in Z value over the change in horizontal distance. This ratio is sometimes called rise over run. In fact, you calculate slope by dividing rise by run. If you're using elevation, you might detect that over a distance of 8 miles, the elevation has increased or decreased (depending on your direction) by 40 feet. The equation (rise/run) would be 40 feet divided by 8 miles. If you divide 40 by 8, you get a slope or gradient of 5 feet per mile (5 ft/mi).

You can also calculate slope by percent slope. To do this calculation, you need to have the numerator and the denominator in the same units. So, for example, a change in elevation of 200 feet (rise) for every 1,000 feet of distance (run) gives you 2/10 or 0.2. If you multiply 0.2 by 100, you get a 20-percent

slope. If you want to measure your slope in degrees, apply fancy inverse trigonometric functions. But most GIS applications usually rely on the two basic methods I outline. If you want to see all three of these methods side by side, you might visit

```
http://geology.isu.edu/geostac/
        Field_Exercise/topomaps/
        slope_calc.htm
```

✔ **Orientation:** You usually hear orientation compared to compass direction. Good compasses are normally calibrated to 360 degrees, where 0 and 360 are the north direction. Not all GIS software follows this method of determining direction (for example, 0 might be east rather than north), nor does it all calibrate in 360-degree segments. Some GIS start numbering clockwise from east (0), and some reduce the number of segments from 360 to a smaller set of numbers to save computer space and speed up calculations.

The direction in which a surface changes (*orientation*) is also very telling. For population data, it tells you where population is increasing and where it's decreasing. Barometric surface orientation tells you the direction of the wind. The orientation of depth to ore body data gives mining engineers an idea of where the ore body is large and where it's small. You can find many possible applications of surface-change directions.

Working with Surface Data

The idea of a geographic surface means that categories of geographic features have Z values (for a third dimension) that occur in different places. Z values are recorded data, called *statistics*. Statistical surfaces have three properties:

- ✔ **They contain Z values distributed across geographic space.** Because you know that your data are distributed throughout the surface, you can predict values in places where you don't have sample data.

- ✔ **They're measured and recorded at interval or ratio scales.** Because your data are interval or ratio scale numbers, you have really precise statistical numbers to analyze, rather than just names or "big, medium, and small" ordinal measurements.

- ✔ **You can assume that they're distributed continuously.** A continuous surface means that data points exist everywhere and that you can predict them. But if you have a discrete dataset, such as population by county across the lower 48 states, you can still do some of this.

These three factors give you serious power to manipulate the statistics to predict missing values and group data based on your needs. The following sections show you how to sample and analyze the statistical surface. And, like always, rules are made to be ignored. You get the scoop on which rules you can ignore, and when and how to ignore them, in the section "Ignoring the rules," later in this chapter.

Collecting surface data for entire areas

Topographic surface data are usually collected at point locations with survey instruments and GPS units. Sometimes, however, data are collected for a whole area all at the same time. If you want to show a map of population across the lower 48 states of the United States, you can't possibly know where each person is standing at any given time, so you collect data by some political subdivisions, such as counties. So, each county gets only one number.

You can map both of these datasets (topography or population by county). You can even make the discrete data collected by county look like data collected at point locations. I explain the difference in the section "Understanding Discrete and Continuous Surfaces," earlier in this chapter.

Sampling statistical surfaces

Continuous statistical surfaces require sampling because the data occur everywhere. You can sample a surface by following these steps:

1. **Examine what the surface looks like.**

   Take a look at the highs and lows of your surface, and get a feel for its overall structure.

2. **Locate where the surface is smooth and where it is rugged.**

   Smooth surfaces have little change in Z value over distance while rugged surfaces change rapidly over distance.

3. **Define a *sampling strategy* (how many samples to take and where to take them) based on your understanding of the surface change.**

   For example, setting a strategy based on surface change may mean that you take more samples for rugged surfaces.

Sampling surface data requires you to know something about the surface. A general guide is that the more rugged the surface, the more samples you need to take because a rugged surface has more changes in Z value than a smooth surface. Smooth surfaces have less information (defined as change in Z value over horizontal surface) than rugged ones. When you want to represent a smooth surface, you don't need to take as many samples to capture the information as with a rugged surface.

Taking a lot of samples of a surface doesn't necessarily increase the quality of your map. This fact is called the *point of diminishing returns.* Computer programs don't actually need all the data to perform their interpolations, and you can reach a point of diminishing returns in which you gain little or no benefit by adding more sample points. The software might not even use all the additional points you provide. Not to mention, taking more samples requires more effort, which typically costs money. All that overhead, and you may not even improve the quality of your map!

Sampling statistical surfaces requires more than just knowledge of where the information content is the highest (that is, where there are rapid changes in Z value with distance), although that knowledge is extremely important.

To define statistical surfaces, you don't need to provide continuous data although there is still the assumption that they are continuous. In some circumstances, data are collected by region, rather than at specific locations. For example, you can sample census data by census tract or census block, housing costs by neighborhood, or voter survey data by county. Because these types of samples aren't linked to specific locations on the Earth, you need to find a place to locate them if you hope to use them for analysis.

Your GIS software may require you to put a sample location inside each polygon if you plan on doing any surface analysis — especially if you plan to predict missing values based on the data from the surrounding areas you've sampled.

Interpolation is the technical term for predicting the Z values of an area based on samples that you've collected for other areas. For example, if you know that you're standing on a gentle slope at an elevation of 100 feet and your friend, who's 400 feet away from you, is standing at an elevation of 200 feet, you can predict the elevation of someone standing halfway between you and your friend as 150 feet without actually having to go halfway and measure.

For discrete surfaces, you can use two basic ways to locate points for interpolation:

✔ **By using the centroid (geometric center) of each polygon:** For a rectangular polygon, the centroid is right at the intersection of two lines drawn diagonally from each corner.

✔ **By using the center of gravity of the distribution:** The center of gravity method means that you pick a point that's closest to the location of the highest concentration of your samples.

Figure 13-5 shows the centroid of a polygon and center of gravity of a distribution. Suppose the polygon in the figure is a county, and you want to record the number of people who voted in that county. You can assign that number to a point in the county — for example, the centroid point could represent the number of voters. But if you know that most of the people in the county live on the south side, you can move the recording point toward the bottom of the polygon (to the center of gravity) so that it more accurately represents the distribution.

The center of gravity method of interpolation is much more accurate than the centroid method, but it also takes more time and requires that you know the locations of all your samples. In many cases, you don't have this information. If you do have all the known locations, the GIS places the point at the center of gravity. For most applications, especially in which you're working with small polygons, you can adequately represent the data collected for that polygon by using the centroid.

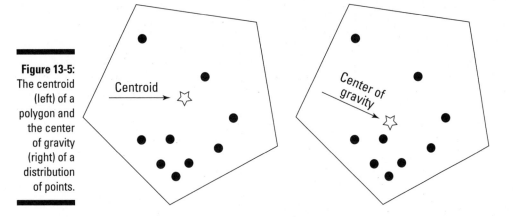

Figure 13-5:
The centroid
(left) of a
polygon and
the center
of gravity
(right) of a
distribution
of points.

Displaying and analyzing Z values

Continuously distributed Z values measured at the interval and ratio scales
provide a wide array of both display and analytical capabilities. You can find
displays (basically, anything that shows the output graphically — usually as a
map) useful for visualizing the nature of the surfaces at a glance, and even for
comparing some other maps to the surface by using a process called draping.

In *draping,* the GIS literally drapes layers over the statistical surface. You
can use this process when you want to see, for example, the relationships
between vegetation and elevation values. This draping function is visual and
doesn't necessarily produce a new map, but when you see common patterns,
those patterns may suggest a map overlay that you can use to analyze them.
For information on the techniques of map overlay, see Chapter 17.

Without the need for overlay, statistical surfaces allow you to analyze the
Z-value data themselves. Continuous data provide information about changes
in Z value from place to place, directions in those changes, and the rapidity
with which the change takes place in different portions of your map. With
some forms of continuous data, you can determine the movements of air,
water, populations, wildlife, diseases, and many other features. Other forms
of data allow you to determine whether portions of your study area are visible from an observation point. You can define watersheds and use them
for modeling hydrological conditions and deciding where to build bridges. I
show you the most important of these techniques related to non-topographic
surfaces in the section "Predicting Values with Interpolation," later in this
chapter. In Chapter 15, I show you some of the really powerful tools available
for topographic statistical surfaces.

Ignoring the rules

You can work with data that are distributed continuously, rather than discretely, much more effectively when analyzing statistical surfaces. You can have difficulties when using non-continuous surfaces because the values change suddenly — and not always predictably. Surfaces allow you to predict changes, which is one of surfaces' most useful properties. You might like to analyze many distributions, such as those related to population, and economic factors so that you can present them as continuous surface displays and interpolate missing values.

Let me tell you a little secret: If you look carefully at the description of statistical surface properties in the section "Working with Surface Data," earlier in this chapter, you notice it says, "You can assume that they're distributed continuously." That part of the description tells you that the data can be "assumed" to be continuous because, if you suspend your disbelief (pretend they are distributed continuously) for just a bit and wink at the data, you can literally ignore the more rigid part of the definition that says they must be continuous. So, you can display and analyze maps of population distributions (discrete data) just as if people really did occur everywhere, as shown in Figure 13-6. This assumption of continuity doesn't cause a problem because even truly continuous surfaces are composed of a discrete subset of sample points, rather than a potentially infinite number of points.

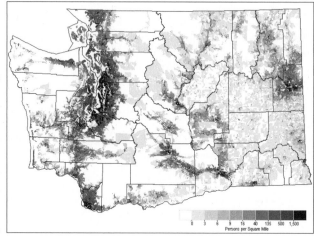

Figure 13-6:
Using a gradient display to map discrete population data as a continuous surface.

0   3   6   9   16   40   135   500   1,500
Persons per Square Mile

Predicting Values with Interpolation

You can easily figure out the word *interpolation* if you just look at the first part — *in*. When you *interpolate,* you try to predict or estimate missing values *in-side* a set of numbers by using numbers on either side of the gap. Interpolation is based on the idea that numbers occur in a predictable fashion or sequence. You could fill in this sequence easily — 10, 20, 30, ??, 50, 60. You come up with the value 40 because the numbers don't occur randomly; they exhibit a sequence (each number is 10 more than the previous number).

Surfaces' values change in sequences. These sequences determine the way you predict the missing value. Some sequences for surfaces are smooth and steady, and others are rough and undulating. Some sequence of numbers can represent each of these types of surfaces, but the sequence might be very complex. Some sequences might be linear (such as 1,2,3...) and some quite non-linear (for example, *logarithmic* — rapidly increasing or decreasing curves). The interpolation method you choose depends on the nature of the surface, as described in the following sections.

Determining values with linear interpolation

Linear interpolation is a method of determining missing values that assumes the surface you're examining changes in a regular, arithmetic progression — such as 1, 2, 3, 4, 5, and so on. Few — if any — natural surfaces satisfy the arithmetic progression. Still, it provides the basis for all the other methods. Essentially, all methods of interpolation are modifications of the linear method based on a combination of both different progressions (number series, such as logarithmic) and advanced knowledge of the surface (knowing how Z values tend to change over distance).

Linear interpolation is simplicity itself. You place a group of elevation values collected by survey or GPS on the map in their exact X and Y locations. From there, you can use linear interpolation to determine the elevation at specific intervals between two endpoints. Follow these steps:

1. **Determine the elevation for each endpoint.**

   In this example, the elevation values are 200 feet at point A and 400 feet at point B, as shown in Figure 13-7.

2. **Measure the horizontal distance between the endpoints.**

   For this example, the distance between points A and B is 1,000 feet.

3. **Determine the total change in elevation between the endpoints.**

The surface elevation changes by 200 feet (400 feet minus 200 feet) along a distance of 1,000 feet. You assume that the surface elevation for this distance changes in a uniform arithmetic fashion.

4. **Specify uniform intervals between the endpoints.**

The surface elevation changes by 20 feet for every distance of 100 feet. So, you just need to divide your distance between the two points (A and B) into ten 100-foot chunks.

5. **Figure the elevation for the point located at any specified interval.**

A point that's three 100-foot chunks (intervals) away from endpoint A has an elevation of 260 feet, according to this calculation:

```
starting elevation + (number of intervals × elevation
        change per interval) =
        requested elevation
```

For this example, you're looking for the elevation at the third interval from endpoint A:

```
200 feet + (3 intervals × 20 feet per interval) = 260
        feet
```

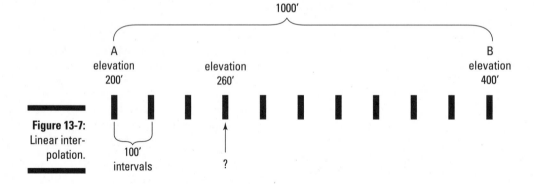

Figure 13-7:
Linear inter-
polation.

Using non-linear interpolation

Incline ramps are a classic example of a linearly changing surface. I've seen many natural areas that are very flat and have a gradual surface incline, but even the most uniform of these areas has imperfections and places where the surface changes in a different way than the majority of the surface. Because you can't assume that natural surfaces are linear, you need to find methods of interpolation that are non-linear.

Non-linear methods of surface interpolation try to match the numeric progressions they use to the progressions that most closely approximate how the surface changes. To most effectively decide which of these non-linear interpolation methods to employ, you need a close understanding of the nature of the surface, including both

✔ A conceptual idea of how the surface looks before you map it

✔ A working knowledge of what factors influence how the surface might react to changes in other factors.

When selecting a non-linear interpolation method, first characterize your surface (gradual, steep, or a combination; smooth, rugged, or some combination). Then, become familiar with the nature of the processes at work that cause the surface to exist. With this knowledge, you can match the techniques that best suit the way the surface changes.

Estimating values with distance-weighted interpolation

The weighted-based-on-distance method is more properly called *inverse distance weighting*. The closer things are to each other, the more likely that they're related. So, if you have a bunch of numbers (say from GPS data) that you can use to interpolate missing data, the numbers that are closest to the missing data probably have more meaning. This method gives a higher weight to closer values and less weight to those that are farther away. This type of interpolation method (weighted) is one of a growing number of methods called *exact interpolation methods*.

Here are some examples of how distance to something affects its importance for interpolation:

✔ If you need to purchase a single candy bar, do you go to the local convenience store just down the street or travel the 8 miles to the supermarket, where you might save a nickel or a dime?

✔ Do you talk to your friends who live 1,000 miles away as much as you talk to the people who live in your immediate neighborhood?

✔ Your favorite music group is having a concert next week 500 miles away, but the same group will be in your city next month. Are you more likely to go to the concert at home or 500 miles away?

You can always find exceptions to these examples, but generally, the closer you are to your candy bar, your friends, or your favorite rock band, the more likely you are to interact with them.

The same idea applies to surface data. Consider temperature data. If you're standing at one place where the temperature is 75 degrees Fahrenheit and you take a step to your left, you expect that the temperature is pretty close to 75 degrees in that location, as well. But if you get in your car and travel 100 miles from your original location, you probably don't expect to find the same temperature at your destination.

Knowing the other exact interpolation methods

You use exact interpolation when you can't take the risk of assuming a linear change in Z value. The word *exact* means that many parameters are taken into account, such as the relationships of distance to weight of the interpolation. Some exact interpolation methods require the user to define search radiuses. Other interpolation methods don't require the user to define radiuses. Some interpolation methods calculate the point at which distance has no effect on interpolation (meaning that no relationship apparently exists between the surface values that are so far apart).

You can find a comprehensive list of interpolation methods here:

```
www.spatialanalysisonline.com/output/html/
          Comparisonofsamplegriddingandinterpolationmethods.html
```

The table that you find at this site lists the interpolation method, speed, type, and comments that can assist you in deciding which of these methods to use.

I find a particular method — called *polynomial regression,* or more commonly, *trend surface* — quite useful for making surface generalizations. This method doesn't provide an exact prediction of missing values but deviates from exactness to give a general idea of how the slope is trending. You can use this kind of trend surface method if you just want a general idea of a surface — and the accompanying general trend in a distribution — rather than all the gory details.

Chapter 14

Exploring Topographical Surfaces

· ·

In This Chapter

▶ Using topography to identify what you can and can't see

▶ Working with stream basins

▶ Understanding the relationship of topography to flow

▶ Identifying the parts of a stream

· ·

*T*opography deals with the configuration of the Earth's physical surface and landforms, their characteristics and locations, and their graphic depiction on maps. Topographic surfaces can be flat and uniform, or deeply dissected by streams and their associated *watersheds* or *drainage basins* (areas that drain in the stream). Some topographic surfaces are gentle and rolling, moving fluids slowly, and others are rough and steep, causing fluids to move rapidly with great erosive force.

You can find a staggering number of forces that create diverse land surfaces, as well as different surface types. Fortunately, earth scientists have grouped, summarized, and generalized most of these forces and their resulting effects to produce some pretty substantial models that you can use. In this chapter, I show you three basic groups of models based on topographic surfaces: viewshed analysis, basin analysis, and stream *morphometry* (stream shape analysis).

Modeling Visibility with Viewsheds

Analyzing terrain so that you can identify places that are (and aren't) visible from a known location on the earth is called *viewshed analysis*. Just like a watershed is a place where you can find the presence of water, a *viewshed* is a place where you can observe an object in your line of sight. You can analyze visibility from two perspectives:

✔ **Point to point:** Analyze visibility from one point to another point. You have to know the location of your *target point* (the place you do or don't want to see) and the location of the *observation point* where you (the observer) are standing.

✔ **Multiple locations to multiple locations:** What if you aren't staying in one place? For example, say that you plan to walk along a mountain trail and you want to make sure that — during your walk — you don't see the housing development in the area. Unlike point to point analysis, you determine the visibility of multiple locations along the path to multiple parts of a housing development. Same problem, same solution — just more computing.

The importance of viewshed analysis

You can use the ability to determine what you can and can't see from selected locations in many ways. Military planners use this technology so that they can place their troops and equipment in places that are hidden from as many locations as possible. Some telecommunications systems require line-of-sight analysis so that they can properly place their transmission equipment. In urban areas, city planners like to hide the unsightly highways from residents' view; viewshed analysis helps determine where to apply terrain and even vegetation surfaces to solve the problem. In many cases, obscuring the view also produces a quieter environment. For example, Department of Transportation (DOT) workers erect barriers that both obscure and quiet the sound of large highway systems.

Viewshed analysis can tell users where they can see things and where they can hide things from view. If you subtract from the map the area that you can see, you're left with the area that you can't see. One area is the complement of the other. Here's a list of some additional possible ways that you might use viewshed analysis:

✔ You want to build a resort retreat that's obscured from the city but provides a mountain view.

✔ You're putting up billboards along a curvy highway, and you want motorists to see the billboards for the longest time possible.

✔ You need to know where to place cellular towers so that as many people as possible get good reception.

✔ You want to know the best place to put your military forward observers so that they get the best possible view of the battlefield.

✔ You need to set up fire watch stations that see both over the topography and over the trees.

You may find viewshed analysis helpful in any number of applications. Just ask yourself these questions:

✔ Will the terrain affect whether I can see certain features?

✔ Do I want to see those features or not?

Then, let the GIS work its magic.

Using ray tracing

When you look at topography, you can easily imagine a line drawn from each of your eyes to each point in the landscape they view. These lines are identical to the lines that physicists work with in the study of *optical geometry* (the graphical portrayal of the properties and behavior of light from place to place using lines to show how light interacts with objects it encounters). They draw lines from a subject's eyes, through lenses, and to and from viewed objects. GIS viewshed analysis uses the same technique. The GIS draws lines from the observer to the observed. GIS users call this technique *ray tracing.* The GIS software is literally tracing a line from your observer to different locations on the map. Figure 14-1 shows ray tracing in action. Notice how the line drawn from the observer to a point on the surface shows places that are visible and those that are hidden from sight (shaded).

Figure 14-1:
Simulating line of sight with ray tracing.

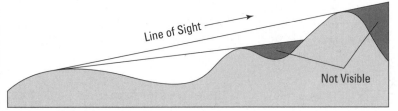

You can use ray tracing to analyze one line of sight at a time, but you typically assess visibility for all possible locations everywhere within the map. By comparing one or more observer locations with all other locations, the software defines the visible and hidden polygons from a potential observer. You can examine visibility over a surface by creating a complete viewshed, such as the one in Figure 14-2, but you may also want to make this even more realistic.

The simple approach assumes that you, the observer, are on the ground and that you're searching your topography to see what you can see. This approach usually employs only a single layer to analyze, and that layer is normally a topographic layer such as a Digital Elevation Model (DEM). The real world is a bit more complicated than this model would suggest, unfortunately. Real landscapes have trees, buildings, and other obstructions. Plus, when you observe terrain, you don't normally lie flat on the ground with your eyes at ground level.

You can easily build these little surface features into your viewshed model. If you're the observer and you're standing, your eyes are probably anywhere from 4 to 6 feet off the ground, depending on your height and the shoes you're wearing. So, if you're working with a topographic model and have no vegetation or buildings to get in the way, you need to adjust only for the height of your eyes from the ground. To make that adjustment, you just add that height value — say, 5 feet — to the observer's elevation. If you're observing from a tall building or a lighthouse, add that additional height to the existing elevation at your observer's location. You also account for obstructions such as trees in the exact same way. Just add their heights to their locations' elevations. Figure 14-3 shows the line of sight for an elevated observer and or when obstructions block the way.

The most accurate viewshed models add the heights of obstructions from data in other map layers, as well as from the observer's height. Most of the time, though, you don't know the observer's height, and your elevation model isn't accurate enough for these minor improvements to make a huge difference. Use these refinements only when you have very good elevation data and a thorough knowledge of the observer and the obstructions.

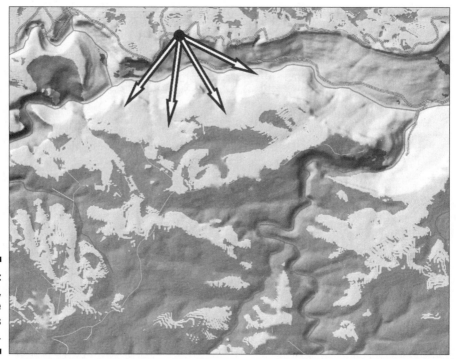

Figure 14-2:
A viewshed, where the light areas are visible.

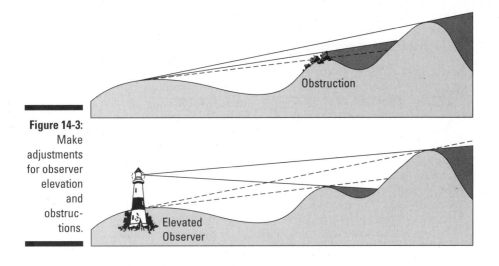

Figure 14-3:
Make adjustments for observer elevation and obstructions.

Finding and Using Basins

You can find analyzing stream configuration useful for characterizing the flows along these networks (as I talk about in Chapter 15), but you can do much more with the terrain models inside your GIS. For example, you can analyze what the watershed or basin within which your streams reside actually looks like. The basin determines the nature of your stream, and the basin also controls water accumulation (which I discuss in the section "Working with basins in your GIS," later in this chapter).

These three metrics — stream ordering, basin analysis, and water accumulation — are all inextricably linked to one another. You can't really do stream ordering until you can define the stream based on a thorough knowledge of the basin. When you have a complete model of the basin, you can finally model the movement and accumulation of water.

Knowing how basins work

Think about what happens in a basin, whether a washbasin or a stream basin. When water enters the basin from above, that water immediately begins to seek its way downhill because of gravity. Water acts differently, depending on whether a basin is steep or shallow. Steep basins tend to have really sharp cuts in them because the water is moving so much faster than it does in a shallow basin and therefore cuts deeply into the surface. The software must define the movement of water based on the shape of the basin. You usually begin with a set of elevation data from some source, such as a Digital Elevation Model (DEM).

Basins or watersheds are composed of the area upslope from the stream network in such a way that all the water landing on the watershed could potentially flow overland into that stream network.

Working with basins in your GIS

To define both streams and the basins in which they reside, you must first develop a grid (set of grid cells) that illustrates where the water would accumulate if you filled the basin with water. The *accumulation* is really a measure of distance based on where the water will go. The deeper the basin, the deeper the potential water accumulation. These deep areas are near the central trunk stream (the large stream in the center). Shallow parts of the basin are typically near the tributaries (the branches from the central stream) because water accumulates through movement from tributary streams and from overland flow.

Like with all topographic surfaces, a basin (a low landform associated with a stream network) can include many little depressions and peaks. The GIS can model these differences but the process is unnecessarily complicated. In this discussion, I pretend that my surface has no minor variations. This *depressionless surface* allows for the development of a general model. You can most effectively model stream basins by using a raster GIS, as I do in this section.

After building your topographic database, you can create a model of the grid surface. The highest grid cells — the ones on the outside margin of your basin — define the basin's boundaries. The section "Finding and quantifying streams," later in this chapter, discusses how the lowest grid cells for each sub-basin (a part of the basin associated with a tributary stream) form the cells that make up the stream network. The places where the stream grid cells form each sub-basin are called *pour points* because, at these locations, water pours from one sub-basin to another.

You want to identify the location of the watershed boundaries mainly so that you can model the accumulation of water while it flows downslope into the basin itself. While you move from the highest elevations, each grid cell contributes water to the total water moving downslope. Because each grid cell downslope receives the water from the grid cell above, the values accumulate.

The more grid cells upslope, the larger the values at the bottom (as shown in Figure 14-4). The left part of the diagram shows the elevation values, the right shows the direction the flow will move based on those values. The GIS records these directions as a series of codes (shown at the bottom of Figure 14-4) based on direction. So, for example, a flow to the right would be coded a value of 1, a flow to the left would be 16, and so on.

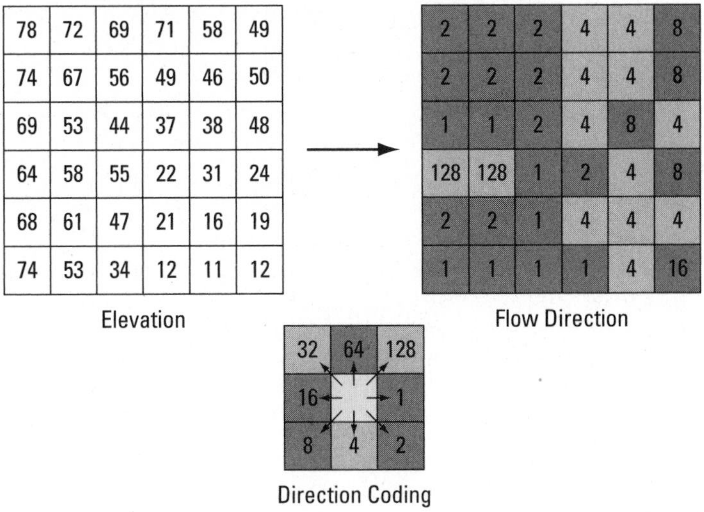

78	72	69	71	58	49
74	67	56	49	46	50
69	53	44	37	38	48
64	58	55	22	31	24
68	61	47	21	16	19
74	53	34	12	11	12

Elevation

2	2	2	4	4	8
2	2	2	4	4	8
1	1	2	4	8	4
128	128	1	2	4	8
2	2	1	4	4	4
1	1	1	1	4	16

Flow Direction

Direction Coding

Figure 14-4: Basin accumulation.

In this simplified version of how the accumulation model works, each grid cell usually has additional attributes that include values such as the amount of precipitation, and absorption rate of the surface over which the water moves. These values are similar to the friction values that you use when you measure functional distance, as discussed in Chapter 12. The GIS can include these values for more sophisticated modeling operations by adding both the cells and the friction values while the accumulation proceeds.

Characterizing Flow

Watersheds are pretty irregular in elevation values, even if you ignore the occasional highs and lows in the surface. These irregularities result in water moving in different directions. When you add differential surfaces to your GIS database, you can build some really powerful models of flow direction. The following sections show you the basics of how flow direction works.

Knowing the importance of flow

The direction of the movement of liquids downhill determines where erosion is likely to take place, how quickly the water will move, and the moving water's possible destructive force if important structures happen to be in its path. People build many structures in floodplains and basins. While water moves down slopes, it not only increases in speed and force, it also saturates surfaces, dislodges portions of those surfaces, and picks up debris that contribute to the potential for damage.

Modeling and using flow

To determine flow direction, you need a grid of elevation values, such as a Digital Elevation Model (DEM). You evaluate each grid cell by comparing its elevation value to that of its neighbors. In Figure 14-5, the GIS compares each of the outside grid-cell elevation values to the center (target) cell. After it determines which cell has the highest drop in elevation, it assigns that cell a code that indicates the direction of water movement.

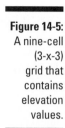

Figure 14-5:
A nine-cell
(3-x-3)
grid that
contains
elevation
values.

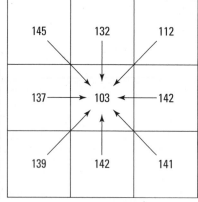

In Figure 14-5, the input grid shows an elevation value for each of the nine grid cells. By using the center grid cell's value as the starting point (essentially, where the water is at that time) and comparing it to each of its eight neighbors, the computer tries to find elevation values lower than the starting grid cell (in this case, the center). If the elevation values are higher than the starting cell, the computer moves to values that are lower. GIS software packages handle similar surrounding values differently but normally perform a search beyond the immediate neighborhood of cells. The GIS software looks for the *direction of steepest descent,* meaning the one neighboring grid cell that has the steepest drop in elevation from the starting cell. In Figure 14-5, the grid cell in the upper-left corner has the largest elevation drop to the central grid cell of all the eight neighbors.

To let the computer know that the largest elevation drop is in the upper-left corner, you need to give the flow a direction value. Depending on your software, the eight possible direction values could start on the top, bottom, left, or right. The ESRI software uses the output values shown in Figure 14-6 for direction. Direction coding starts at the right (east) with a value of 1 and moves clockwise, doubling the value it uses for each sequential grid cell. This scheme was decided for programming reasons (that you don't need to worry about), so just realize that the coding scheme reduces the number of direction numbers from 360 compass directions to only a handful.

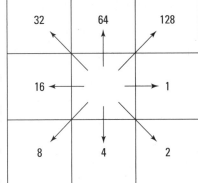

Figure 14-6:
Output
values for
coding the
direction of
flow.

You may want to know which way the water is going to flow, but if you also want to get a feel for how fast that water will move, you need to measure the distance it will cover. This general model for flow is calculated with a very simple formula:

Flow = Change in Z value/(distance × 100)

On a 3-x-3 grid, a GIS compares each beginning (center) cell's elevation value to each of its neighbor cells. The GIS also measures distance in this way:

✔ Cells that share a side with the center cell have a distance of 1.

✔ Cells diagonal to the center cell have a distance of 1.414.

Some strange situations can occur when the computer determines flow direction and distance. For example, all the neighbor grid-cell elevation values may be higher than the starting cell, or several grid cells may all have the lowest elevation drop. Different GIS packages have different ways of dealing with these issues, including some unique coding schemes and (my favorite) allowing the program to expand its search beyond the initial eight grid cells to find an appropriate drain point. I like this expanded-search approach because it allows the program to evaluate elevation values that are more likely to be different than those immediate eight cells.

Defining Streams

The stream is a very common land feature. Streams usually don't occur in isolation. They usually have a lot of branching tributaries. The numbers and positions of streams have an impact on stream flow, water accumulation from overland flow, and even the ecological conditions of the streams and the upland corridor (often called a *riparian corridor*).

In vector GIS, you can find streams basically by selecting the appropriate line entities based on their labels. After you know the locations of the streams, you can measure the complexity and configuration of the network so that you can understand the processes at work in that network. Defining streams in a raster GIS takes a bit of work, which I outline in the following sections.

Finding and quantifying streams

Raster GIS contains some of the most powerful analysis related to streams and their landforms because of the raster data model's capacity to easily represent every piece of the topographic surface. To perform *stream morphometry* (a way to model how tributary streams connect to bigger streams to make stream networks), you first need to define the locations of the streams and their *watersheds* (areas that drain into the stream, as shown in Figure 14-7). These locations are related to elevation. Streams mark the low spots in the drainage basin area, and all the parts of the watershed must be at higher elevations than the adjacent part of the stream network so that the water will drain into the streams.

Ecologists like to be able to model stream connections because those connections provide clues about the biological organisms that exist both on the land and in the streams. *Geomorphologists* (scientists who study landforms) use this information to understand how the streams will change over time through the processes of erosion and deposition. Most GIS software includes the well-established methods of stream morphometry.

To determine where the streams are in a watershed, use topographic characteristics that define how the stream works by following these basic steps:

1. **Define the outside edge of the basin and its sub-basins as the locally highest locations in the map.**

2. **Define the *pour points* (the low points) for each sub-basin.**

   Each pour point forms a junction or branching for the stream.

3. **Calculate the least elevation surface values upward from the pour point to find the locations of the streams**

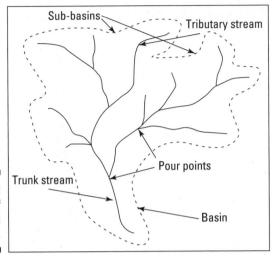

Figure 14-7:
Parts of
a stream
watershed.

The stream branchings at the pour points allow you to determine stream morphometry by providing the skeleton of your basin. When you calculate accumulation and define basins (as discussed in the section "Finding and Using Basins," earlier in this chapter), you identify the stream network. A GIS assigns grid cell values to reflect water accumulation (which I talk about in the section "Characterizing Flow," earlier in this chapter). The grid cells with the highest values, called the *stream cells,* represent the actual stream. You can reassign these high values to be any number you want so that you can identify where the stream is located. To make these cells stand out even more, most GIS software assigns all the other background (non-stream) cells a value of NODATA, which tells the computer that the background cells have no numbers assigned to them at all.

I like to assign numbers to stream cells that are wildly different from any values that might occur anywhere else in the database. For example, if you know that your non-stream cells' values won't be bigger than 100 during the accumulation calculations, you can assign a value of 200 for stream cells. You can always renumber the stream cells after you classify all the background cells as NODATA.

After locating the stream, you need to perform a stream-order analysis. Sometimes called the *bifurcation ratio* because the streams tend to break in twos, the *stream-order analysis* calculates the relative locations and connections of a stream network, characterized by the order in which the streams come together. First-order streams, for example, are commonly the smallest outside tributaries whose water comes only from overland flow and not from other streams.

Identifying methods that work for you

You can calculate stream order in several ways, but the two most common are the Strahler method and the Shreve method. Both of these methods use the exterior link as the first-order stream link and assign these stream tributaries a value of 1 (representing first order), as shown in Figure 14-8:

- **Strahler method:** The most common method. Each new stream order is based on the convergence of two streams that have identical stream-order numbers. When dissimilar stream orders come together, the computer assigns the higher of the two tributaries' orders to the next stream tributary.

  For example, you can use the Strahler method for studies involving sediment movement and stream flow analysis.

- **Shreve method:** Adds the stream orders together to produce the next stream-order number, thus accounting for all possible stream-order combinations. You call these numbers *stream magnitudes,* rather than orders, because the values aren't based solely on the ordering.

  Ecologists use the Shreve method to reflect how the different stream linkages often result in considerably different ecological conditions.

You can use stream order to define buffer distances because a GIS codes these numbers right into the links themselves.

Figure 14-8:
The Strahler and Shreve methods of determining stream order.

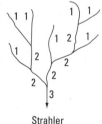 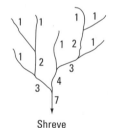

Strahler   Shreve

Some software uses different methods to determine stream ordering — but most of those different methods are really variations on a theme. The Strahler and Shreve methods are the most common approaches to this type of stream drainage analysis.

Before you start doing work with stream-network analysis, check your software documentation. It usually describes in detail how the methods work and may even provide suggestions for selecting a method.

Chapter 15

Working with Networks

. . . .

. . . .

*T*he Earth is filled with linear objects. You can find treelines, fault zones, shorelines, fences, rows of houses — the list goes on. Some linear objects are special not so much because of what they are, but how they work. These special linear objects are called *networks*. Networks are collections of connected linear objects such as roads, railroads, or rivers that branch from place to place. They come in different sizes, numbers of branchings, and angular configurations. One classic type of network is the stream network, which allows movement of water along its length. A network that has this movement along its length is called a *corridor*. This chapter covers the different properties of corridors and shows you how to measure them.

Measuring Connectivity

You can measure connectivity in a network by comparing the number of actual node-to-node links that exist in a given network to the maximum number of nodes that are possible. This measure of connectivity is called the *gamma index*. Usually, the index ranges from a value of 0 (which indicates no connected links at all) to 1 (where all possible links are connected). Both general and transportation-specific GIS software packages contain algorithms that allow you to calculate the gamma index, which the software records as decimal values.

Recognizing the importance of connectivity

The gamma index gives you a way to determine whether the network is as connected as possible. I once used this index to model the speed at which edge species (species that like to live on the edges of different types of environments) would inhabit a network based on whether that network was poorly, moderately, or highly connected. But most applications of the gamma index relate to human transportation systems.

If you look at a road map of the *conterminous* United States (the lower 48 states), you may notice that the central portion of the country (the Great Plains) has far fewer road-to-town connections than do the coasts. Most of the coastal areas are highly connected so that each town has many connected links. In other words, the coasts have a gamma index that's approaching 1. In the Great Plains, many towns aren't connected directly to some nearby roads, forcing you to take longer routes to get from place to place. Thus the gamma index for road networks in the Great Plains would be closer to 0.

Measuring and using connectivity

A GIS calculates the gamma index by counting all possible connections (links) among all the nodes in the database and then dividing this total (the number of actual links) by the maximum number of possible links. You use the gamma index connectivity figure indirectly when you perform shortest-path analysis (covered in the section "Finding the shortest path," later in this chapter), but it can also be useful for

- **Characterizing the loss of shelterbelt networks:** Shelterbelt *networks* are rows of planted, fast growing trees that you use primarily to protect crops from the negative effects of wind. By evaluating the gamma index of shelterbelts, you can begin to plan which regions need the most urgent replantings of trees to replace those that are dead.

- **Evaluating increases in road network connections over time:** Check out Figure 15-1, where the dots represent towns and the connecting lines represent roads. From left to right, the three networks illustrate how the number of connections among the towns increases while time passes and more roads are built. This increase in connectivity may indicate increased urbanization.

- **Analyzing the complexity and efficiency of utility networks:** Like roads, utility networks provide for flows, but the flows are of electricity, gas, or water. The more connected the utilities are, the more homes they can serve. Some utilities, such as electrical service, often require that substations be appropriately placed to ensure that, if one goes down, others can still provide electricity to all the customers. Analysis to determine appropriate placement is essentially one of connectivity.

Figure 15-1:
The gamma index increases while connectivity increases.

Working with Impedance Values

When you treat networks as corridors, the movement of fluids, objects, animals, and/or vehicles is your primary concern. For any movement, the nature of the network as a corridor has an impact on how fast things move and in some cases, even whether they move at all. The resistance to movement is called *impedance* and can be a function of the size of pipeline in a water delivery system, the roughness of dirt or gravel roads, or the number of lanes in highways. The most common use of network impedance is to model street patterns for transportation routing. I use that example in the following sections to show you how network impedance works.

Knowing when your paths are fast or slow

Street patterns provide an excellent example of how factors can affect speed in a network. Speed may increase or decrease for several reasons. Here's a quick list:

- ✔ Posted speed limits help determine the maximum travel speeds for streets and highways (sometimes even the minimum speeds).

- ✔ Fluctuations in traffic density often determine how fast the traffic will move at different times of the day.

- ✔ Accidents and other unplanned disruptions are bound to slow traffic, especially if lanes are closed off.

- ✔ Street repairs and changes in road condition often result in changes in traffic flow.

- ✔ The existence of border checks, sobriety checkpoints, weigh stations, and other regulatory facilities result in slower traffic flow.

- ✔ Changes in weather, especially the existence of dust storms, rain- and hailstorms, blizzards, and sleet storms will often have huge impacts on traffic flow.

Modeling impedance for traffic flow

Given the many factors that affect the movement of flow, especially traffic flow on streets, modeling these changes really improves your ability to know when your grocery deliveries will arrive, how fast a fire engine can get to a fire, the fastest way to get to a particular restaurant for dinner, or the route that you might need to take to pick up your children from their many activities.

When you model impedance, you do have to deal with one complexity — most software gives you many options. And you simply won't use some of those options because they require that you know much more about your traffic network than you probably do. Ever see those little black traffic counters (they look like rather large power cords) running across your streets from time to time? Traffic engineers and transportation geographers use these counters to determine how much traffic flows past a particular point at any given time. You could easily include this important information in your GIS as impedance data. Other common traffic-flow impedance data include speed limits, stoplights, and stop signs.

Here's a short list of some of the attributes, or options, that you can put into a GIS impedance layer that you might use for traffic modeling (this list omits data related to turns, which I discuss in the section "Working with Turns and Intersections," later in this chapter):

- ✔ **Impedance attribute:** How long it normally takes to travel a certain distance.

- ✔ **Default cutoff value:** The value at which the computer stops searching for a location. You can override or change this value.

- ✔ **Accumulation information:** A list of possible attributes that accumulate while distance increases, including costs, students riding a bus, and many more.

- ✔ **Restrictions:** Restrictions that you place on the use of links (for example, the permitted types of traffic on portions of your road network). For example, you can choose to force hazardous cargo to use certain streets.

- ✔ **Hierarchy:** A set of rules for travel, regardless of how your software determines the impedance values.

These factors (and others that depend on the software you use) are sometimes referred to as the *Origin-Destination* (OD) matrix. The data for these factors form a separate table. When you use the GIS for transportation modeling, the software compares the network by examining the OD matrix so that it knows the rules to apply and where to apply them. The OD matrix contains several components, including origin, destination, and barrier information.

By reading the OD matrix, the software knows the ID codes you have placed at strategic locations, how to read the impedances you have linked to those ID codes, and other necessary factors that give you the power to simulate how real traffic flows operate.

 Use your knowledge of traffic patterns and impedances to decide which actual values to include in the OD matrix. For example, your matrix entries will reflect attributes such as speed limits for roads, maximum flow capacity for water in a pipeline network, or the amount of electrical resistance you might get from a particular type of wire in an electrical utility network.

Working with One-Way Paths

Impedances, like surface friction values, can be low or high, short distances or long, and even prevent movement entirely. Dead ends, checkpoints, and bridge collapses prevent the movement of travel. You model these impedances by placing no-movement commands at these points. However, you can restrict movement only to one direction by creating *unidirectional (or one-way) paths.*

Understanding unidirectional paths

In certain situations, travel along a network normally takes place only in a given direction. A fast-moving stream moves rafts in only one direction. Some gate entries allow traffic to move in only one way and out another for control — some gates even place those nasty spikes to rip up your tires if you try to back up. The most common example of a unidirectional path is the one-way street. I focus on that example in the following section because it's easy to understand and modeled frequently in transportation GIS applications.

Modeling unidirectional paths

If you want to move ambulances, fire engines, or even grocery trucks, you usually look at a map to figure out the best way to get from point A to point B. More than one argument between friends or spouses has begun over disagreements about which path to use. Perhaps you should purchase some GIS software for your friends and loved ones to reduce these disagreements. How would GIS software help? The software is pretty adept at handling the factors that influence travel through a network. One of the most common (and aggravating) situations that street maps don't often show is the location and direction of one-way streets.

One-way streets are coded in a GIS by including a travel direction restriction in your OD matrix. When the computer tries to model vehicle access to a north-directed one-way street from the south, or even to make turns into it in the opposite direction, the GIS stops the movement and redirects the search to look elsewhere for a path. In essence, the effect is the same as when you put up a barricade across the street that prevents travel in that direction (see Figure 15-2).

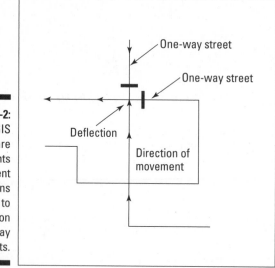

Figure 15-2:
The GIS software prevents movement in directions opposite to the direction of one-way streets.

Characterizing Circuitry

A network's complexity is partly measured by the numbers of links it has compared to the maximum number it could have (the *gamma index*). One other factor both characterizes the complexity of a network and adds to the robustness of traffic modeling capabilities. That characteristic, called *circuitry*, is based on the idea of closed loops. Closed loops allow moving objects, fluids, and so forth to travel alternate routes when moving along a network (see Figure 15-3). The following sections show how closed loops work.

Knowing when lines create circuits

If you're modeling streams, you may encounter a location where a stream splits around an island and can move in two different directions. In forests, some wildlife takes advantage of circuits to avoid predators because the prey identifies more than one path that they can use to escape. British-designed road systems provide another classic example of circuits with their intersection roundabouts. In the United States, the interstate bypass system allows you to travel around major urban areas without dealing with slower intra-urban traffic.

When you perform traffic analysis for routing, allocating customers or bus riders, or finding the nearest facility, having circuits greatly increases your route options. In some cases, the routes may be longer but could save travel time because of reduced traffic and less traffic-related impedances.

Measuring and modeling circuits

The algorithm for modeling circuitry is very similar to the one you use for network connectivity (which I discuss in the earlier section "Measuring and using connectivity"). Rather than a gamma index, you use another ratio called the *alpha index*. In transportation geography, the alpha index is used to characterize the amount of circuitry. The alpha index, like the gamma index, is a ratio. In the alpha index, the ratio compares the actual number of *circuits* (loops in a network composed of nodes and links) divided by the maximum circuits possible. Like with the gamma index, both general and transportation-specific GIS software has the alpha index built in for you.

Although you may find having an index of circuitry handy to characterize areas, GIS users more commonly use networks for traffic-pattern analysis. You treat the loop just like any other connection inside the GIS. So, these loops provide alternatives for movement, and you don't have to do any additional work to set them up or model with them.

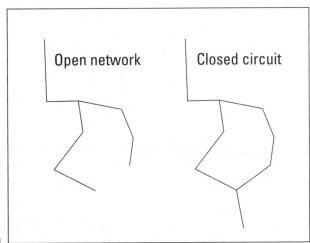

Figure 15-3:
A circuit;
also called a
closed loop.

Open network   Closed circuit

Working with Turns and Intersections

Whether your network database has circuits or just different sets of branches, you sometimes must turn or cross another branch that intersects your path. You have to make some turns at right angles, some at acute angles, and some at obtuse angles. You may even encounter a link that meets a dead end or a T that offers two possible directions. These different types of linkage provide an array of alternative movements that the GIS can handle.

Recognizing the importance of turns and intersections

When you plan a trip, you need to know where you're starting, where you're going, what the traffic will probably be like, which roads or streets you must take, and where you turn from street to street. Some routes may require you to travel more distance than other routes. Some routes have stoplights that slow your movement. Although your map may not show it, some streets are one-way streets — and those streets might lead to your destination or might flow one way in the wrong direction. Turning from two-way streets to one-way streets means that the turns themselves also restrict your movement by forcing you into specific lanes on which you can travel.

Stoplights, stop signs, and different types of street intersections all either add or detract from your travels. Planning a trip means more than just looking at the lines on a map and tracing the shortest route. One time-saving trick I like to use is to plan a trip where I make only right-hand turns. Experience has taught me that cutting across traffic (with left-hand turns) can be difficult and time-consuming because you must wait for an opening to get across.

Because you no longer have to rely on a static, hardcopy road map, you can include knowledge about what factors speed travel and what factors slow it down when you travel along a road network. The more compete this information, and the more accurately it represents how intersections and turns work, the more likely you can accurately model traffic. The key to successful traffic network modeling, then, is getting all the right information into the computer.

Encoding and using turns and intersections

You can code turns for any junction where edges connect. The number of edges connecting at each junction limits the number of possible turns. The maximum number of turns is the number of edges n squared, or n^2. You can even model a single edge where your maximum number of turns is one, $(1^2 = 1)$. Figure 15-4 shows graphically how this formula works with a T intersection composed of three links.

When you have three links (or edges), squaring that number tells you that you have nine possible movements, including U turns and straight-through (non-turn) movements, as follows:

- ✓ **Three U turns:** Each edge provides an opportunity to do a U-turn if you allow it.

- ✓ **Two right turns and two left turns:** The link coming down from the top allows you to make right turns along the inside lane, and left turns to the outside lane.

- ✓ **Two non-turns:** The links on the bottom allow you to go straight through, or turn into the vertical link from the right or the left, depending on your approach.

Turns are defined by rules that affect conditions in which two edges come together. Every GIS employs some set of rules that govern how and where you can make turns. These rules will likely reflect the terminology used in your GIS software package, but, in general, GIS turn rules give you these restrictions:

- ✓ You can't set up turns that include impossible movements or seriously violate traffic laws.

- ✓ You can create U-turns, but you can turn that ability off if, for example, U-turns aren't legal.

Some software stores turn information as network datasets, and others use turn tables that include a field for adding a turn impedance (see the section "Working with Impedance Values," earlier in this chapter). Turn impedance values, which network datasets also include, give you the opportunity to include the amount of time it would normally take to perform a turn with traffic or against it. So, you can model vehicular movement very realistically.

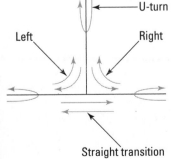

Figure 15-4:
The possible turns in a T intersection.

Directing Traffic and Exploiting Networks

Networks provide a selection of really useful techniques that you can employ in businesses, government, organizations, emergency services, and many other applications. After you add to your GIS a complete network that has appropriate road designations and ID codes for your impedance values, the impedance values themselves, plus turn information, you have one of the most powerful networking tools available anywhere. Here's a quick list of the operations you can use in most GIS software:

- Finding the best route (shortest, fastest, or even the most scenic)
- Finding the closest geographic feature to you
- Finding service areas, such as what parts of town are served by a single fire station

Finding the shortest path, or route

Many people use street maps to figure out the shortest route from their current location to some specified location. If you search for the shortest path, you use the length of the network links to decide the best route. Because you're not concerned with time, you don't need to include impedance values in your search. Although the GIS ignores most impedance values when you specify that they aren't needed, the software always includes some indication of one-way streets and dead ends so that it doesn't send you on a wild goose chase.

Most GIS use several algorithms to determine the shortest path, but most of these algorithms look at a number of alternate routes and choose the shortest. One common method of shortest path search is called the *best first* algorithm, which takes the shortest path available at each step. Other algorithms try a number of alternatives at predetermined distances and select the best. Consult your software documentation for more detail, but most of these techniques provide pretty good results.

Even if you don't own a complete professional GIS, you may have already seen this ability to find the shortest path in action. Yahoo! Maps and Google Maps find the shortest path from one town to another or one address to another. These packages are rudimentary GIS software and do a reasonable job for general routing tasks. For detailed, time-critical routing, especially for business and emergency services, these packages typically don't include the detail that you need to efficiently navigate through road networks, especially during critical times of the day like rush hour.

Both the rudimentary packages and the more sophisticated GIS software provide two forms of output. The first form is a map that shows a highlighted version of the route. You can expand this map to a turn-by-turn map that includes a text version of exactly how far each leg of the route is, exactly where the turns are, and which direction you turn (like the example in Figure 15-5).

The shortest path means shortest in terms of distance traveled. If you want to get there fast, use the fastest-route feature of your GIS (see the next section for more on this feature).

Finding the shortest route to one destination is pretty handy, but you can also use the GIS to find the closest of a group of possible destinations, as well. You've probably used this kind of feature when you wanted to find the closest Sears or Wal-Mart store by using the store's Web site. The software finds the addresses of all the possible locations and performs a shortest-path analysis on all of them from your current location. Businesses have found that providing interactive maps directing traffic to their front door (so to speak) is a really powerful marketing tool.

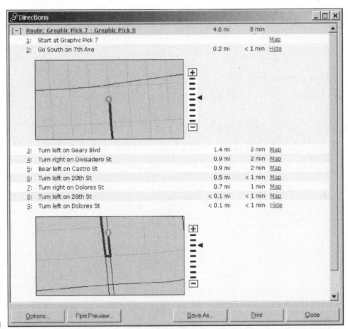

Figure 15-5:
A turn-by-turn map with the directions.

Finding the fastest path

The shortest way from one place to another isn't always the fastest, as shown in Figure 15-6. For example, you might be able to get from the southeast corner to the northwest corner of a big city much faster by taking the interstate around the city, rather than taking the shorter diagonal route. The shorter diagonal route includes slow speed limits, traffic congestion, stop signs, stoplights, and a host of other possible impedances.

The fastest-path search allows your GIS to start from one location and accumulate impedance values while it searches through the network to find your destination. Instead of keeping track of just distance, it combines distance and impedance to calculate the route that has the least amount of total impedance. Typically, GIS software measures impedance in time, so a street that has a 25 mile-per-hour speed limit has a higher impedance value (longer time) than one that has a 75 mile-per-hour speed limit.

You can use the fastest-route search method for another useful search. If you know where you're starting but don't know your destination, the GIS can find your destination for you. Okay, the GIS isn't going to find the perfect picnic spot or the best restaurant unless you give it a clue. Perhaps you have a sick family member and you need to get him or her to the hospital. If your town has four or five hospitals, the GIS can search for the closest one. You need to tell the software that you want it to find the fastest path to all the selected hospitals in your area and the fastest of those is the one you would select.

Figure 15-6:
The shortest path versus the fastest path.

▪ ▪ ▪ ▪ ▪ ▪ ▪ Fastest Path   ▬▬▬▬▬▬ Shortest Path

One aspect of the fastest-path approach is a bit unique because your impedance values are based not just on speed limits, impedances, and typical traffic volumes, but also on the time of day. Some construction zones operate at different times of day, traffic congestion is worse during rush hours, and even the season might impact which roads are open. Encode in the OD matrix the times of day when traffic is congested on which streets to make sure that the software can determine how drivers can avoid those streets at those times. In some hilly cities and towns, selected streets are blocked off at times in winter because the snow and ice make those streets dangerous to drive. You can code all sorts of impedances as windows of time along selected portions of the network and store them in an attribute table, as shown in Figure 15-7.

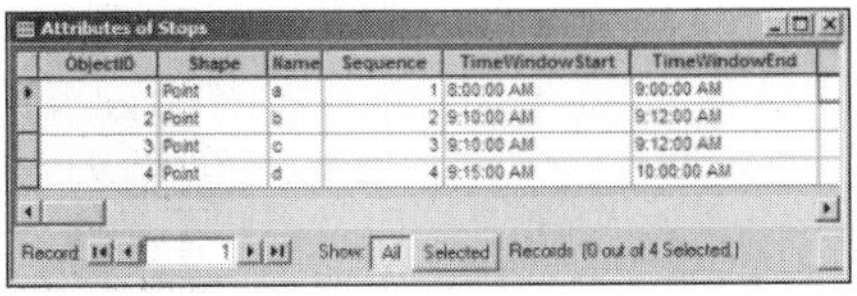

Figure 15-7:
Setting
up a time
window
table in the
ESRI ArcGIS
network
module.

Finding the nicest path

You may be concerned about factors other than speed or distance. Many highways have scenic routes that are a bit slower or perhaps even longer than a more direct route, but you may want to view the scenery. GIS software can help you find the scenic routes, as well as the fastest and most direct routes, because you can create scenic impedance values. When you run your search, you simply ask the GIS to search for links that have the nicest scenery. You just need to remember to include the values that you want to search for in your network database attributes.

Finding the service areas

The network database allows you to create all sorts of values to put into your tables. You can include distance, impedance, turning, and even scenic information. One really powerful tool for service providers and businesses alike is the ability to find service areas.

A service area identifies places that are within a reasonable distance for use and relies on knowing how many people or houses are located along individual links in a network, much like impedance values, except that the database has to know how many people are along each link. The software starts at some point location and searches either along a chosen path or outward in all directions. While it searches, it counts all the houses or people it passes until it reaches the end of the database, some specified search radius, or some maximum quantity of people or houses.

You may find the service area tool useful in a number of different settings. Here's a list of some common uses:

- ✔ Creating bus routes for school children, limiting the search by the number of passenger seats on the bus

- ✔ Determining how many fire stations a town needs, based on a search of the number of homes one fire station can safely service (based on historical numbers of fires and other factors) in a worst-case scenario

- ✔ Marketing to potential newspaper subscribers based on an accumulated search of households that don't currently subscribe to your paper

- ✔ Creating service areas in which a pizza delivery store can deliver pizzas within an assigned time period

Figure 15-8 shows the results of a search that returns all areas in a portion of a city that a driver can reach within five minutes. Service-area analysis allows you to use all the impedance and turning tools that you use for any other analysis. This process of finding service areas, often called *allocation* by geographers, is a very popular tool among business analysts because they can not only allocate for their own facility, but also compare themselves to their competition. Allocation is one of the more popular economic placement tools available in GIS today.

Figure 15-8:
The results of service-area analysis showing areas that are reachable within five minutes.

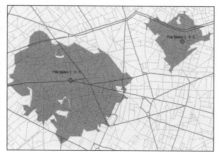

Chapter 16

Comparing Multiple Maps

. . . .

. . . .

Geographers have been aware for generations that spatial patterns sometimes repeat themselves. The usefulness of comparing these spatial patterns has been documented as far back as the Revolutionary War when General George Washington's French cartographer made multiple hinged maps that showed the locations of both British and American armies in the corresponding geographic area. This idea probably goes back as far as maps themselves.

In the early 1960s, the idea of using computers to make maps brought with it a renewed interest in comparing multiple map patterns. Landscape architects used clear acetate to physically overlay and compare maps manually with a fair degree of success. Urban and regional planners found comparing different maps essential to the observation, quantification, explanation, and eventual exploitation of multiple patterns.

Even before map overlay came into use, people observed patterns occurring in the data contained on various maps for an area. Auto thefts happen in locations that have cars, shopping happens in areas that have stores, roads connect towns, vegetation responds to different soil types and different slopes, cities mostly occur near water, certain wildlife prefer selected vegetation, people of similar income tend to live in the same areas, and so on. You can observe each of these coincident patterns both in the field and on maps. You can analyze the degree to which the patterns correspond, and figure out whether cause and effect relationships exist, much more easily if you can overlay the maps.

This chapter shows you the three main ways to compare multiple maps. I show you how you can use mathematics and formal logic to make the comparisons. You see how overlay works in raster and vector data models, and how to use selective overlay to compare particular features across maps.

Exploring Methods of Map Overlay

The ability to overlay maps was a primary driving force during the early development of GIS software. Initially, overlay functions focused on comparing one set of polygons to another, but they quickly expanded to comparing polygons with point and line features. For people, overlaying one map on another is a pretty simple concept involving a visual and intellectual interpretation process. But getting a computer to do the overlay is a bit more complicated because the interpretation process can take several forms and may even combine different methods. Fortunately, you don't need to worry about these details because the methods have been formalized into some very structured approaches in the GIS software.

All professional GIS software has some form of overlay toolkit that allows you to choose among different methods of overlay and select the layers you want to use. Most GIS toolkits have a separate graphical user interface or offer a button, icon, or tool to click that, in turn, gives you a lot of choices. Check your user manual or type **help overlay** in your software's help utility.

Knowing the overlay methods available helps you find the right tools in your own software. Map overlay operations compare three pairings of geographic feature types:

- Points to polygons
- Lines to polygons
- Polygons to polygons

The first two pairings, points to polygons and lines to polygons, use a *presence/absence* overlay method. In other words, either the points or lines coincide with (occur in or run through) a particular type of polygon or they don't. This method is actually a little more complicated than that (as I explain in the following sections), but not much.

The last pairing — polygon to polygon — introduces some additional potential methods of comparison that, to some degree or another, simulate the intellectual process of visual map overlay. Each method is adapted to either raster or vector GIS data models, as I describe in the following sections.

Finding points in polygons

Point features occur in many different places at different times for different reasons. You might want to identify where and to what extent (for example, how many) selected point objects occur within the boundaries of area features. And guess what? You can do just that with ease by using either a raster or vector GIS model. Table 16-1 shows a selection of questions that you might ask to identify whether features you're interested in occur in the same locations.

Table 16-1	Possible Point in Polygon Pairings	
Question	*Point Feature*	*Polygon Feature*
Do the nests of the endangered burrowing owl occur within the boundaries of the city parks?	Burrowing owl nests	City parks
Is there a difference in number of auto thefts in the Hillshire neighborhood than other parts of town?	Reported auto thefts	City neighborhoods
Which parts of town have the highest concentrations of cancer patients?	Homes of cancer patients	City neighborhoods
Are cases of salmonella distributed evenly among the counties of the United States?	Cases of salmonella	U.S. counties
Do brown bears use habitat types differentially (use one type more than another)?	Brown bear sightings	Wildlife habitat areas
Which counties have the greatest concentrations of tornado sightings?	Tornado sightings	U.S. counties
Do certain parts of town demonstrate low numbers of newspaper subscribers that I can target for new subscriptions?	Homes of newspaper subscribers	City neighborhoods

You can probably think of a bunch more questions that deal specifically with your business or industry, but I think you get the idea. To answer your questions, you need a piece of software that can keep track of the data you use to make your comparisons.

The GIS (by design) keeps track of point features, their X and Y coordinates, and their attribute information. It also stores the coordinate and attribute information for all the polygons in the locations covered by your database. When you put data into your GIS database, you usually put point and polygon features on separate layers (although you don't absolutely have to). So, to compare these different features, you retrieve the point features (by type or value; see Chapter 11) from one layer and compare the result to polygons — also selectively retrieved by type or value — from another layer.

This process is called *point-in-polygon overlay,* and you follow these general steps to accomplish it in your GIS:

1. **Use your system's built-in query capabilities to find point features by any attribute you decide.**

   The retrieved points are your first layer in this instance.

2. **Use your system's built-in query capabilities to find the polygon features that you want to compare with the point features.**

   The retrieved polygons are your second layer in this instance.

3. **Place the point and polygon layers on top of each to create a new map.**

   Your GIS creates a new visual display that shows both the points and the polygons. It also creates a table that combines the attribute information from each layer.

Creating a new map from your separate layers is important when you want to actually use that map for analysis. Unlike a strictly visual overlay that you might see in a graphics program or many CAD systems, GIS software relates not only the locations you see (the X and Y coordinates), but also the attribute information for every feature you're comparing.

You use the tables that you get when you perform point-in-polygon overlays to analyze the results. These tables indicate which points occur in which polygons (as shown in Figure 16-1). Using this information, you can count the number of features for each polygon, calculate density by polygon, and perform a host of other analyses.

My discussion of point-in-polygons overlay is based on the vector data model, but you can easily do the same thing by using a raster GIS. When using a raster GIS, the software searches the attributes in the same way that it does in vector and compares the coordinates — but the exact X and Y coordinates are replaced by the grid cell's position in the matrix.

The point-in-polygon operation is a powerful tool for comparing point objects to area objects, but you might not need to perform this operation if the points and polygons are already on the same map.

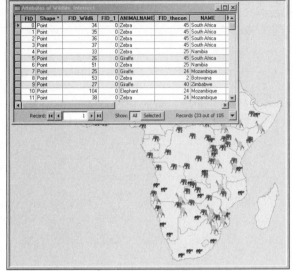

Figure 16-1:
A table
resulting
from a point-
in-polygon
overlay
operation.

Finding lines on polygons

The movie title *A River Runs through It* pretty accurately explains the idea
of *line-on-polygon* overlay methods. I changed the word *in* (from the point-
in-polygon method I describe in the preceding section) to *on* in this case
because the vast majority of lines don't fit completely inside any given polygon. Instead, like a river, they cross polygon boundaries. (Perhaps using *line-
through-polygon* would be more appropriate, but few people use that term in
the GIS industry.)

Lines running through polygons may be just as meaningful for analysis as
points in polygons. In many applications, such as transportation planning, the
line features are the primary focus of the analysis. Transportation planning,
for example, includes many types of lines whose presence can have profound
impacts on the surrounding polygon features. Table 16-2 shows some examples of line-on-polygon comparisons.

Table 16-2	When Lines and Polygons Collide	
Line Features	*Running through These Polygons*	*Have This Effect*
Fault lines	Residential areas	Produce populations at risk
Streams and irri-gation systems	Agricultural lands	Irrigate crops

(continued)

Table 16-2 *(continued)*

Line Features	Running through These Polygons	Have This Effect
Hazardous cargo routes	Hospital zones	Create potential evacuation difficulties
Paths and roads	Portions of national parks and preserves	Provide access for eco-tourism and conservation activities
Water pipes, electrical lines, and gas lines	New subdivisions	Determine the cost of infra-structure development
Tree lines and fence rows	Farms and grazing lands	Provide protection from the wind or livestock containment

In vector GIS, lines are composed of line segments, each with its own set of beginning and ending coordinate pairs. Most of the time, these beginning and ending locations don't coincide exactly with the borders of polygons. So, the software has to work a little harder to figure out exactly where the line moves from one polygon to another by determining where the line is going and where it will intersect the coordinates of the boundary for each new polygon (as shown in Figure 16-2). Both raster and vector GIS models can perform line-on-polygon operations, but GIS users more commonly use the vector model for this operation.

Like with point-in-polygon methods (discussed in the preceding section), this overlay operation produces tables that show which lines and polygons intersect so that you can analyze properties like orientation, density, and comparisons of line and polygon types.

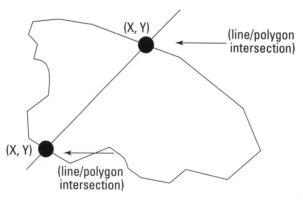

Figure 16-2:
Finding the coordinates for points where a line crosses a polygon's borders.

Using Logical Overlay to Compare Polygons

The most powerful, most robust, and probably most common types of overlay operations are those comparing one set of polygons to another. Generally known as *polygon overlay,* they offer many more options than point-in-polygon and line-on-polygon overlays.

Logical overlay is a method of comparing multiple maps that uses a group of operations based on *set theory* to search the polygons to determine whether attributes are shared from one map to the other (belong to a common set).

In set theory, three basic set operations compare the objects contained in one set to those of another set. These operations are *union* (where you combine all the stuff from both sets), *intersection* (where you select only those things common to both sets), and *complement* or *symmetrical difference* (where you identify all the objects that the sets don't have in common).

Table 16-3 shows some possible uses of polygon overlay.

Table 16-3	Polygon Overlay Uses
Polygons	*Comparison*
Land uses or land cover	Time periods
Plant and animal species ranges	Common distributions
A time series of maps of Dutch elm disease	The rate of the spread of Dutch elm disease
Neighborhoods	Which fall within a 100-year flood zone

Unlike comparing sets of dinnerware or sets of keys to make decisions, with GIS software, you look at portions of each polygon that you overlay to see how those portions relate to other polygons in different layers. Like set theory itself, these overlays employ three basic types of operation. You'll also find some less familiar but quite useful forms in your map comparison activities. I explain each of these forms in the following sections so that you can actually see what they look like.

Searching with union overlay

The first form of logical overlay is called *union overlay* (see Figure 16-3). Union overlay, sometimes called an OR search, collects all the polygons that have any of the attribute search criteria and makes a new map out of them.

So, you might search for all polygons that are classed as agriculture in a land-use map and polygons classed as flat lands in your terrain map. This search returns all lands that are both agriculture and flat. The resulting map would combine these two categories into a larger one. Union overlay maintains the categories from each of the input layers, so you can use them later to perform still more comparisons of flat agricultural land with other layers such as soil nutrient values.

Input   Output

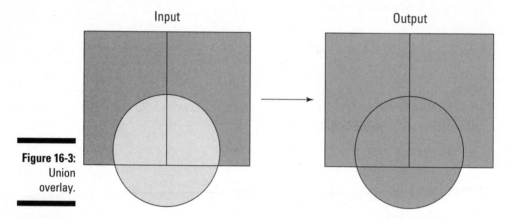

Figure 16-3:
Union
overlay.

Use union overlay when you want to broaden your search and combine different attributes of your polygons. All the polygons that have any of the search criteria are included in the output.

Using intersection overlay

Sometimes, you want to know which polygon features are shared by each of the polygons you're searching. In other words, instead of combining pieces of land for sale with lands that are agricultural, you only want to know which agricultural lands you can buy. You can get this information by using a logical AND search, which needs both categories to exist before it returns a value to the output map. This search typically reduces the size of the area retrieved, instead of increasing it like union overlay does (see Figure 16-4).

Use intersection overlay when you want to narrow your search to satisfy a number of criteria at the same time. This technique can help you find the most appropriate places for some activity.

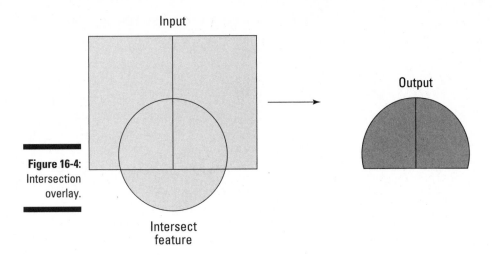

Input

Output

Intersect
feature

Understanding complement or symmetrical difference overlay

Sometimes, when you try to perform a search, you may find it easier to eliminate things that don't match all your criteria, instead of identifying those that do match. This search, called *complement* or *symmetrical difference overlay*, computes the geometric intersections of all polygon types and categories that don't have certain attributes in common (see Figure 16-5). You might want to do this kind of search if you plan to later perform an analysis that relaxes the criteria you originally established.

Input   Output

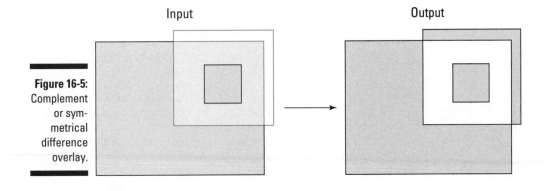

Use complement overlay when you want to know what attributes a group of polygons doesn't share.

Using identity overlay

Identity overlay is similar to intersection overlay, but it treats the layers differently. One layer, the *identity layer,* must be composed of polygons because its attributes determine the identity of everything that coincides with it. But for the other input layer, you can use a point layer (such as wells) or a line layer (such as roads) if you want. When your GIS overlays the two layers, the new shape takes on the attributes of the identity layer (as shown in Figure 16-6). In other words, it takes on the identity of that layer. So, you can have complete control over which layer is more important than another, which is a form of cartographic overlay hierarchy (where one layer takes precedence over another for analysis).

Input   Output

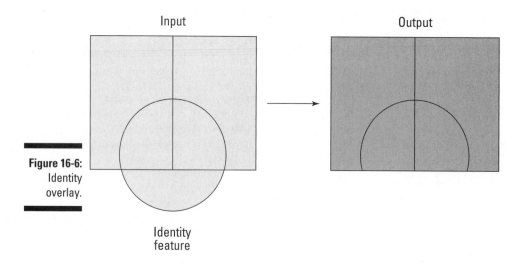

Figure 16-6:
Identity
overlay.

Identity
feature

Use identity overlay when you want to convert intersecting portions of one layer to the attributes of a more important layer. The more important layer is the layer that often controls what values will be assigned to the new polygons, and you usually determine which layer is more important based on what you want to change and what you want to stay the same after overlay.

Comparing geometry with clip overlay

You can use *clip overlay* to compare the geometries of different areas. You simply clip one portion of a map based on the size and configuration of another (see Figure 16-7). Think of clip overlay a bit like a cookie cutter. You might use this overlay if you want to combine information from a small subportion of your study area with the entire study area. For example, if you've done some extensive field research on a small part of a larger forest and want to compare these new data to the original whole-forest layer. Obviously, you need only the small sample area to do your analysis. You don't want the computer to do all that time-consuming graphics work for parts of the database that you don't need.

When you use clip overlay, the attributes of the polygon you clip aren't a factor in determining the size and shape of the resulting layer. Instead, you gather all the information from all layers that you want — but for only a small portion of the map.

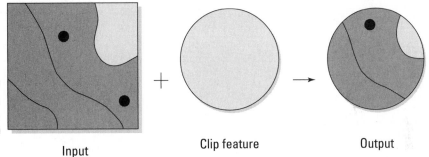

Figure 16-7:
Clip overlay.

Input   Clip feature   Output

I found the clip overlay useful when I was working with a group of researchers studying land-use change during four different time periods. Because we used un-rectified aerial photographs, each of our study areas was slightly different from the others. To fix this discrepancy, we chose the study area that had the smallest footprint (covered the smallest area) as a cookie cutter and clipped all the other layers to match it.

Use the clip overlay when you want two or more layers to have the same *geometry* (size and shape), which allows you to have all the attributes that you want to evaluate, but for a selected portion of the study area.

Understanding Raster Overlay

You can perform pretty much any overlay operation in either raster or vector GIS. In Chapter 17, I discuss map algebra. *Map algebra* is primarily a raster GIS modeling language that allows a huge amount of flexibility in comparing one map to another. Raster overlay is not only more powerful than its vector equivalent, but also less troublesome.

Among the more powerful features of raster overlay is that it can compare raster grid layers by mathematical expression, geometry, and statistical measures; and it's capable of performing a wide variety of operations, including controlling both the search geometry of the overlay and the computational, logical, and statistical methods by which the GIS compares the numbers.

In raster, each grid cell has its own identity. Also, each grid cell's location is directly aligned to the corresponding grid cells on the other layers (co-registered). So, you don't have to worry about all your layers being aligned or problems caused by mismatched edges. Raster overlay also eliminates the need to spend time evaluating very small *(sliver)* polygons to determine whether your GIS created them in error or they really are important.

Use raster overlay when you want to have some serious control over which portions of your map you overlay and what methods you employ.

Comparing Features with Selective Overlay

Selective overlay is not really a specific type or set of overlay operations — it's a unique powerful portion of map algebra normally performed in raster GIS. In many situations, you know exactly which features you want to compare to which others. Frequently, you even want to assign specific new categories and/or numbers to such combinations.

I worked with a team in which we used this technique to examine vegetation at different time periods and compare that vegetation to the landforms. We found over ten types of vegetation and ten types of landforms, making the number of possible combinations rather daunting. But daunting or not, we could more easily decide which combinations we really wanted to examine and do this overlay operation at the same time, rather than do each set as a separate set of overlays. Both approaches would work, but the selective overlay gave us the opportunity to get what we needed in a single operation. It also allowed us to experiment with different sets of combinations quickly.

Suppose that you want to do an overlay of a land-use map from 1974 and another from 2004 that contains soil capability class polygons. You want to decide whether the land uses in 1974 were appropriate by looking at whether those uses persisted through 30 years and worked with the soil capability classes. You can perform a series of individual overlay operations, or you can run the comparison based on a set of rules that you establish. This table includes a few hypothetical examples of combinations that you might want to make with your current scenario.

Input Layer 1	*Input Layer 2*	*Input Layer 3*	*Output Category*
Cornfields 1974	Cornfields 2004	Suitable for agriculture	Sustainable agriculture
Cornfields 1974	Urban 2004	Suitable for urban	High energy use replacement
Cornfields 1974	Abandoned 2004	Unsuitable for agriculture (ponding)	Unsustainable agriculture
Wheat fields 1974	Fallow 2004	Suitable for agriculture	Sustainable agriculture
Abandoned 1974	Cornfields 2004	Suitable for agriculture	Sustainability unclear

Use selective overlay when you have specific combinations that you want to examine but don't want to create a separate overlay for each.

Chapter 17

Map Algebra and Model Building

. . . .

In This Chapter

▶ Using cartographic modeling

▶ Getting the hang of map algebra

▶ Understanding the map algebra language and functions

▶ Creating a model

▶ Going live with your model

▶ Testing to make sure that your model makes sense

. . . .

*I*f you started reading this book here, you're anxious to get started making the GIS work for you. Although retrieving maps and doing all the individual analyses are often enough for everyday tasks, every now and then, you want to watch the GIS spread its wings and do something really spectacular. This chapter explains how you can let your GIS really soar.

The original ideas and concepts underpinning this chapter are the brainchildren of Dr. C. Dana Tomlin and his PhD advisor, Dr. Joseph K. Berry, then at the Yale School of Forestry. Their work has had a lasting and substantial impact on how you can use GIS to make models with maps.

In this chapter, I use the raster model to demonstrate how cartographic modeling works. I also show you the basics of the cartographic modeling language called *map algebra,* the functions that the language provides, and how map algebra creates powerful models. Finally, I show you how you can test the output from map analysis for accuracy and acceptability.

Creating Cartographic Models

When C. Dana Tomlin developed his original Map Analysis Package (MAP) for his dissertation, he wasn't thinking about data structures; he was thinking about modeling. He wanted a tool that would allow him to analyze and combine many maps to produce cartographic models through a process he called (what else) *cartographic modeling.*

Cartographic modeling is an ordered set of map operations designed to simulate some spatial decision-making (making decisions about geographic space) process. The order or sequence of map operations (including creating, combining, and analyzing) is often critical to the success of a cartographic model, and the idea that the result will simulate a spatial decision process is at the heart of a cartographic model. Each map operation follows a carefully thought-out path to model formulation that

- ✓ Identifies each and every output of the individual map operations
- ✓ Outlines each step in how the software derives the output
- ✓ Determines which map layers serve as the information sources with which the software builds the model

People use GIS (cartographic) models only as a spatial decision-making support system. This single application seems to limit the utility of cartographic models, so I like to give the whole idea of decision-making a broader definition. By *decision-making process,* I mean any process that simulates physical or human environments and supports a wide host of possible activities, including

- ✓ **Immediate action:** Activities such as planning, where you decide to immediately employ cartographic modeling for practical resolution of real-world problems.

- ✓ **Theoretical studies:** Constructing cartographic models that simulate natural or human activities in geographic space, contributing to the body of knowledge by describing how spatial distributions of geographic features relate to other patterns or processes.

- ✓ **Future planning:** Information needed for decision-making support can be exploratory (for example, predicting traffic increases at selected locations over the next ten years) and can often lead to eventual decisions and related actions.

You can accomplish cartographic modeling with either a vector or a raster GIS. The operations used in cartographic modeling — map data analysis, and map comparison and combination — don't need a raster data model to work. Even so, the cartographic model is at its most robust in the raster environment.

Understanding Map Algebra

Every modeling toolkit, whether spatial or aspatial (doesn't lend itself to mapping), has a set of individual tools. The carpenter has hammers, saws, nails, and many other tools. The computer programmer has a set of expressions, constants, and syntax rules for each programming language. Every set of tools has a set of procedures that controls how the tools are used effectively. After all, you might be able to pound a nail in with a pipe wrench, but you probably get better results if you use a hammer.

The GIS modeling toolkit is quite robust and demands an equally robust set of guidelines for its proper use. Fortunately, Dana Tomlin and his mentor Joe Berry created a set of rules — a modeling language, actually — that continues to provide a powerful structure for cartographic modeling. This language, called *map algebra,* is loosely based on matrix algebra — but map algebra is both simpler and more structured.

The upper-left corner of Figure 17-1 shows what mathematicians call a *matrix of numbers.* To add this 2-x-2 matrix to another matrix, you add the upper-left number from each matrix, the upper-right number from each, and the lower-left and lower-right of each. You get an answer in the form of another matrix (at the upper-right of Figure 17-1).

Matrix algebra (addition)

$$\begin{pmatrix} 2 & -1 \\ 1 & 3 \end{pmatrix} + \begin{pmatrix} 1 & 1 \\ 1 & -1 \end{pmatrix} = \begin{pmatrix} 3 & 0 \\ 2 & 2 \end{pmatrix}$$

Map algebra (addition)

$$\begin{array}{|c|c|}\hline 2 & -1 \\ \hline 1 & 3 \\ \hline \end{array} + \begin{array}{|c|c|}\hline 1 & 1 \\ \hline 1 & -1 \\ \hline \end{array} = \begin{array}{|c|c|}\hline 3 & 0 \\ \hline 2 & 2 \\ \hline \end{array}$$

Figure 17-1: The similarity between matrix algebra and map algebra.

You can do addition and subtraction pretty easily by using matrix math. You compare the locations of the cells one by one. When you need to use more advanced techniques, such as multiplication and division, for matrix algebra, the numbers aren't compared upper right to upper right, lower right to lower right, and so on. When you work with GIS, this lack of consistency doesn't help you calculate map values because the locations on the Earth are fixed to their locations. Map algebra simplifies matrix algebra so that the X and Y locations don't move when you compare the numbers. Figure 17-2 shows how matrix algebra and map algebra differ when it comes to multiplication.

Matrix algebra (multiplication)

$$\begin{pmatrix} 2 & -1 \\ 1 & 3 \end{pmatrix} \times \begin{pmatrix} 1 & 1 \\ 1 & -1 \end{pmatrix} = \begin{pmatrix} 1 & 3 \\ 4 & -2 \end{pmatrix}$$

Map algebra (multiplication)

$$\begin{array}{|c|c|}\hline 2 & -1 \\ \hline 1 & 3 \\ \hline \end{array} \times \begin{array}{|c|c|}\hline 1 & 1 \\ \hline 1 & -1 \\ \hline \end{array} = \begin{array}{|c|c|}\hline 2 & -1 \\ \hline 1 & -3 \\ \hline \end{array}$$

Figure 17-2: The difference between matrix algebra and map algebra.

Tomlin and Berry saw the usefulness of being able to compare grid cell numbers by using all the power of mathematics. When they created the map algebra language, they recognized that the X and Y locations weren't just matrix positions, but positions on the Earth's surface. Each grid cell location for each layer must correspond exactly with its counterpart in all other layers.

The GIS software must also maintain this correspondence between the locations of grid cell values on multiple layers during analysis — each grid cell on one map must be in the same column and row for its corresponding grid cell on another map. Likewise, each operation will need to follow this rule so that the correct grid cells are compared during analysis. Any serious cartographic modeler is very familiar with this modeling language both because it is so powerful and because it has become the industry standard.

The Language of Map Algebra

Map algebra was developed when GIS software typically used the *command-line interface*, where you actually typed words to get the software to perform its tasks rather than click icons and pull-down menus. Although today's GIS typically employs a graphical user interface, map algebra functionality is still tied to the idea of written commands. Map algebra commands have an English-like language structure, with an action verb (what you do to the map), direct objects (the maps that you retrieve), and a prepositional phrase (describing what you create as a result of the action).

A simple, but typical command might read like this:

> Add (action verb) *mymap* (the object input map) to *yourmap* (another object input map) for *ourmap* (the output map)

When executed, this command uses the power of the map algebra mathematics structure. Each grid cell in each location for *mymap* is added to each grid cell in each location for *yourmap* to create an output map with the sum of each of these individual operations found in each corresponding cell location in *ourmap*. This simplified example gives you the basic idea of how map algebra works.

Performing Functions with Map Algebra

You can use map algebra to perform a wide variety of operations that comprise the cartographic modeling process. These operations include issuing commands themselves, determining when and how often commands are issued, determining the nature of the output, and actually performing the mathematics of the map analyses. Although these map algebra commands used to be part of a command-line system, they exist as part of the graphical user interface of the modern GIS. And so, you work through the graphical interface to apply the map algebra and make your cartographic models.

Exercising control

Most of the control you have over cartographic modeling comes from the user interface. A good example of this user interface is the ESRI ArcGIS Spatial Analyst Raster Calculator. This tool allows you to add many types of input to the modeling process, including raster datasets, shapefiles, coverages, tables, constants, and even individual numbers.

This variety gives the software incredible power to shape the model before the software has even done any analysis. The Calculator searches all the data included in the active database and shows the available input data in its search window. You can then query the available data by using the graphical user interface, which you can use much more easily than command-line queries to find what you need.

Besides providing numerous options for input, the Raster Calculator provides a complete set of operators for arithmetic, trigonometric, and logical comparison activities (like most other raster user interfaces do). You can use these operators — which give you access to an extensive set of mathematical manipulations — to create composite groups of modeling tasks called *functions*.

Essentially, an interface such as the Raster Calculator (shown in Figure 17-3) provides a means to write out complete and often very sophisticated map algebra expressions based on built-in mathematical capabilities. Just a few of these built-in capabilities include addition, subtraction, multiplication, division, roots, powers, absolute values, trigonometric functions, geometric functions, greater or less than, remainders, assignments, and formal logic.

Map algebra expressions that give you control over your model building include

- ✔ **Queries:** Find and input data.
- ✔ **Functions:** You choose functions based on how you want to compare and analyze the data.
- ✔ **Operators:** Direct the mathematical manipulation and logical comparisons to drive the functions.

The power and flexibility of the GIS comes in many different forms in which you can put functions together to take advantage of all the mathematics that the grid cells can use.

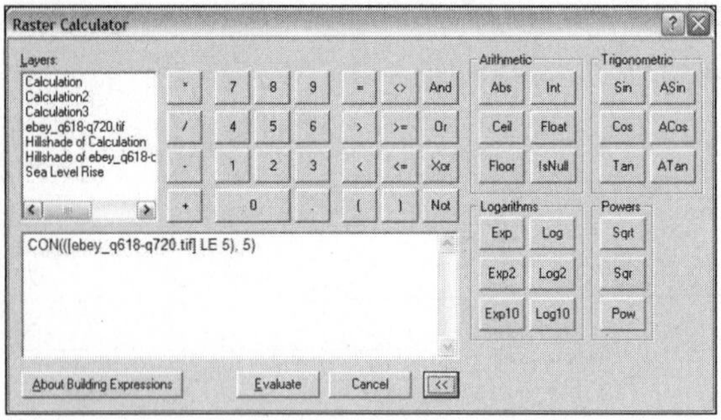

Figure 17-3:
The Raster
Calculator
in action.

Using local functions

The most basic function type is called the *local function* because it operates on each cell individually. Sometimes called by-cell operations, these functions are among the simplest — and most often used — of the map algebra functions. Tomlin used to refer to these functions as a *worm's eye view* of the grid cells because his theoretical worm could see only a single grid cell at a time and so couldn't react with other neighboring cells. You can manipulate each grid cell by using virtually any of the operators, and then compare and combine each grid cell with the corresponding grid cells in other layers.

The local functions are usually grouped into six sets of operators:

✔ **Trigonometric:** Such as sine and cosine

✔ **Exponential and logarithmic:** Such as squares, cubes, and log base 10

✔ **Reclassification:** Such as invert, double, and change all to 1

✔ **Selection:** Such as select all > 2, all between 3 and 10, and so on

✔ **Statistical:** Such as mean, median, and mode

✔ **Other:** Mostly mathematical, such as add, subtract, and conditional comparison

Figure 17-4 shows the input grid and resulting output grid after the software applies a trigonometric sine function, and Figure 17-5 shows a statistical example.

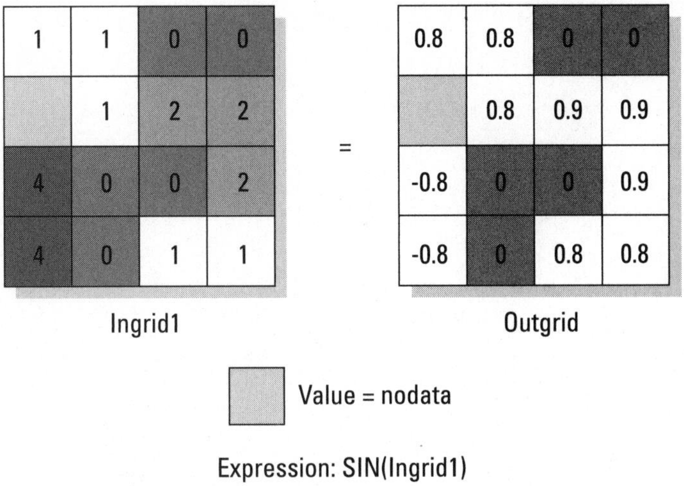

Figure 17-4:
Local trigonometric functions.

Ingrid1 = Outgrid

☐ Value = nodata

Expression: SIN(Ingrid1)

If you understand the basic concept of map algebra, you understand pretty much every local function that you can imagine. Local functions are also easier to explain than those that don't operate on a cell-by-cell basis (such as focal, zonal, and others). You may find the simplicity of these functions handy if you need to explain your GIS modeling to a user.

Using focal functions

Unlike local functions (discussed in the preceding section) — where you zoom in to gain the worm's eye view — focal functions allow you to pull back a bit and begin to see the area around you. This area is sometimes called a *neighborhood. Focal functions* evaluate a single grid cell based on some characteristic of its surrounding cells, or neighborhood. (Focal functions are sometimes called *neighborhood functions.*) In general, the focal function allows you to look at a *target cell* (the cell that you're characterizing or working on), evaluate it based on some relationship it shares with its neighborhood, and return new grid cell value in the same X and Y coordinates but on a new output layer (as shown in Figure 17-6).

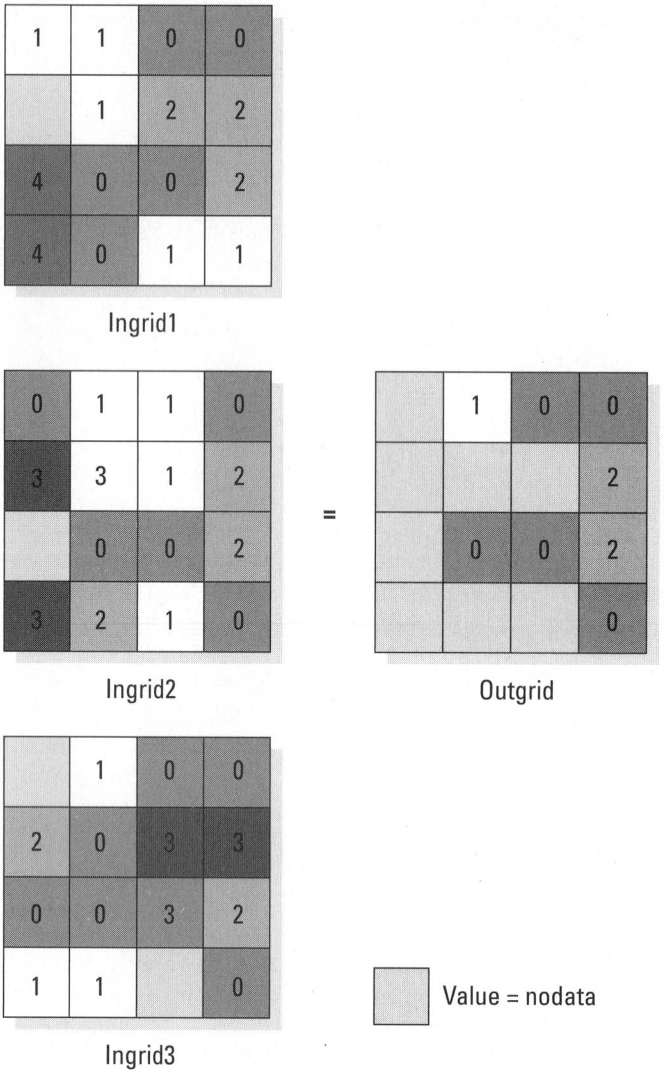

Ingrid1

Ingrid2

=

Outgrid

Ingrid3

Value = nodata

Figure 17-5:
Local
statistical
functions.

Expression: MAJORITY(Ingrid1, Ingrid2, Ingrid3)

Focal neighborhoods are based on a search radius and a shape for the search, often called the *geometry*. Initially, focal functions assume that most neighborhoods include grid cells that border or touch the target grid cell. The neighborhood can, and usually does, extend beyond these bordering cells and can ignore some bordering cells based on the search geometry.

The Neighborhood Function on an Individual Neighborhood

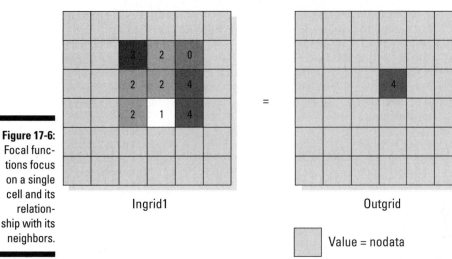

Figure 17-6:
Focal functions focus on a single cell and its relationship with its neighbors.

Ingrid1

Outgrid

Value = nodata

Defining your neighborhood's size and shape

To create a simple search pattern, search for the eight neighboring cells around your target cell in a 3-x-3 matrix. That configuration is called an *annulus* (donut shape) with a search radius of one cell (as shown in Figure 17-7). An annulus could search two cells away from the target, three cells, or more.

The Neighborhood Function on an Individual Neighborhood

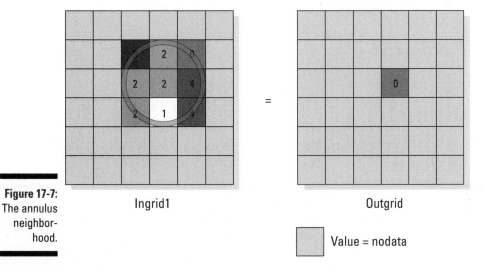

Figure 17-7:
The annulus neighborhood.

Ingrid1

Outgrid

Value = nodata

You can also evaluate neighborhoods that include rectangles of all nine cells (including the target cell), wedges (pie shapes) in different directions, and circles (including the target cell). In short, you can choose from a few different geometries to define your neighborhood search area.

TIP

You select the geometry of your neighborhood for focal functions based on how far away from the target cell you need to go to evaluate grid cells. The distance is often a function of how the real environment you are modeling works (for example, how far from a gas leak you might encounter a fire hazard).

TECHNICAL STUFF

Focusing on focal flow

One focal function is called the *focal flow function* and is really part of the modeling toolkit used for analyzing basins and hydrological conditions (covered in Chapter 14). Focal flow evaluates the neighborhood cells but operates on only the adjacent eight cells — sometimes called the *immediate neighborhood.* Because of this limitation, you don't have to tell the software the search radius or the shape.

In most focal operations, the software returns a value to the focal cell (the target of the calculations). This focal flow operation is sort of backwards because it's really comparing the focal cell to the neighborhood, rather than comparing the neighborhood to the focal cell. You use a focal flow function to determine, given the height of the target cell, which of the outside cells would allow water to drain (flow) into the target (focal cell).

The GIS compares each height of the neighboring grid cells to the height of the target (center) grid cell. If the neighbor cell is higher, it drains into the target cell. The software puts a 1 into that cell. If the neighbor cell is lower than the target cell, the software assigns that cell a 0. If the neighboring-cell and target-cell values are the same, the software uses one of a variety of different ways to indicate no flow. So, this focal function adds complexity and produces eight values, rather than one, because the software returns values to the neighborhood locations, not to the focal-cell location (as shown in the figure).

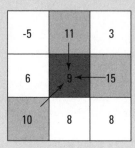

Evaluation for a
single cell location

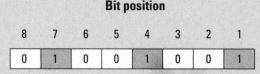

Bit position

Base$_{10}$ bit values for the cell location = 73

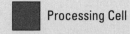
Processing Cell

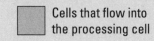
Cells that flow into
the processing cell

You might be scratching your head, wondering how you might need all these different shapes for neighborhoods. Most people will never use all, or even most, of these shapes. Here are a few points about size and shape of impact areas to help you decide on your neighborhood geometries:

- ✔ **The size of your neighborhood area is based on the area of impact of the attributes you're looking for.** Sounds impressive, no? If you know, for example, that people can hear construction noise for ten miles from its point of origin, you can define your neighborhood as a circle with a radius of ten miles.

- ✔ **Circular shapes are common GIS search patterns.** The circle is the most compact two-dimensional shape, and is easy to reproduce. The types of features that occur as circles and can take advantage of this circular search methodology are:

  - Center-pivot irrigation circles

  - Circular animal mounds and ant nests

  - Exposed rock domes and salt domes

  - Some archaeological sites and Indian mounds

  - Coppice dunes

  - Crop circles . . . okay, that's a bit of a stretch — but crop circles exist, no matter who (or what!) creates them

- ✔ **You can search neighborhoods with rectangles because many human features occur in that shape.** You've probably seen many housing subdivisions divided up into square blocks. You could certainly make a block of a certain size into a neighborhood.

- ✔ **Other shapes can simulate real-world shapes during search.** Some shapes are more obscure than circles or rectangles, but some types of features, both natural and man-made, have these kinds of geometries:

  - A ring of vegetation around the base of hills, or around ponds and small lakes, forms an annulus.

  - Wedges might describe alluvial fans and river deltas, and even the pie-shaped crop types within a single center-pivot irrigation feature.

You can no doubt come up with other characteristics that define your neighborhoods. Each feature that you include in your search provides a reason for you to select the size and shape of your neighborhood. The exact syntax for how you make this selection varies with your GIS software, but essentially, you define the shape and the search size as part of your search strategy.

Comparing your neighborhood to the target cell

After you know the size and shape of your neighborhood (as I talk about in the preceding section), you need to decide what properties you want to use to compare your neighborhood cells to the target cell. You can use pretty much all the operators you have available in the local functions for focal functions, as well.

Figure 17-8 shows two simple 3-x-3 annulus examples, one that uses the majority and another that uses the maximum. The first search is a focal majority that returns the number that occurs most often in the annulus to (the eight-cell ring around) the target cell. The focal-maximum search returns the highest number that occurs in the eight-cell ring.

Exploring zonal functions

Unlike focal functions (discussed in the preceding section), zonal functions don't create neighborhoods. They use zones, either from a single grid or from other layers to compare cells by either attribute (description) or shape (geometry). You need to define the zone with which you want to work. A *zone* is equivalent to what geographers call *formal regions. Regions* are areas or groups of areas that share common descriptive information. In a raster data model, a region (or zone) is a group of grid cells that have the same attributes.

Formal geographic regions can be *contiguous* (all in one chunk), *perforated* (with holes that have different attributes), or *fragmented* (like islands that share attributes).

Comparing zonal values

Like with local functions, you can apply a large array of quantitative and logical operators to zonal functions. The most common such operators include majority, minimum, maximum, mean, median, minority, range, standard deviation, sum, and variety.

The general approach to zonal functions follows these basic steps:

1. **Create one input layer, called the *zone layer* (Ingrid 1 in Figure 17-9), as the top layer for comparison.**

   The zone layer defines the zone or zones that you want to evaluate.

2. **Create another input layer that has the values you want to use to evaluate for each zone (Ingrid 2 in Figure 17-9).**

3. **Perform the comparison functions for each zone.**

   For example, you may want to find the variety for each different zone (as shown in Figure 17-10). Finding the variety tells you which zones (Ingrid 1) have the greatest number of different types of features based on the grid categories contained in each zone. Ecologists might use this type of function when they want to identify areas of high biodiversity (variety of habitat, for example).

The Neighborhood Function on an Individual Neighborhood

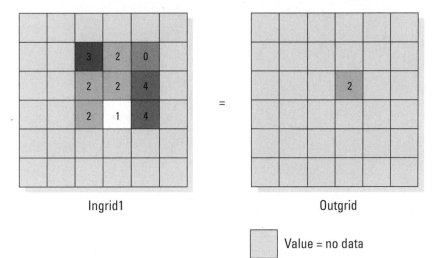

Ingrid1   Outgrid

Value = no data

The Neighborhood Function on an Individual Neighborhood

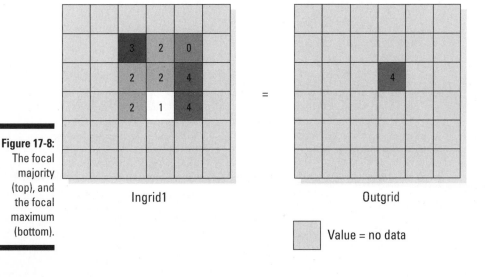

Ingrid1   Outgrid

Value = no data

Figure 17-8: The focal majority (top), and the focal maximum (bottom).

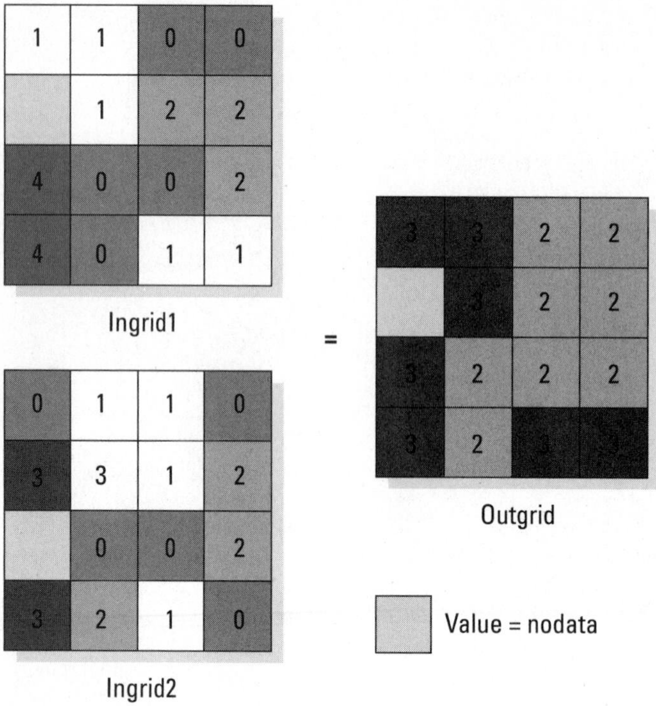

Ingrid1

Ingrid2

=

Outgrid

Value = nodata

Figure 17-9:
Zonal
functions.

Expression: ZONALMAX (Ingrid1, Ingrid2)

Classifying zones based on zonal geometry

Zonal functions augment their analytical techniques by allowing you to classify based on the geometry of the zones themselves. This technique works well for finding individual polygons or groups of polygons of a particular size, as shown in Chapter 10. The most common approaches to classification based on zonal geometry include zonal area, zonal perimeter, and zonal thickness. You may find zonal thickness really useful if you're looking for long and skinny features, for example.

Figure 17-11 shows an example of the zonal area approach, where the software finds and selects zones base on adding up the areas of all the grid cells in each zone. The software then returns the area value to every cell in the zone.

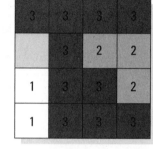

Ingrid1

Figure 17-10:
A zonal
variety.

Ingrid2

Outgrid

Value = no data

In raster GIS, area is calculated by adding up the sizes of all the grid cells included in the zone. Details of calculating sizes of geographic features are outlined in Chapter 12:

✔ **Area:** The number of grid cells multiplied by the area of the ground that each cell represents.

✔ **Perimeter:** The sum of the widths of the outermost grid cells if they're *orthogonal* (edge to edge) and the sum of the diagonals for each cell if they're diagonal.

The zonal function can find the *centroid* (center of an area) for determining the orientation of a polygon (as discussed in Chapter 13). Identifying the centroid allows you to compare the centroid of one fragment or zone to another, which you can use to measure isolation (which I discuss in Chapter 12).

In raster, *polygon* means a collection of grid cells of the same value or attribute.

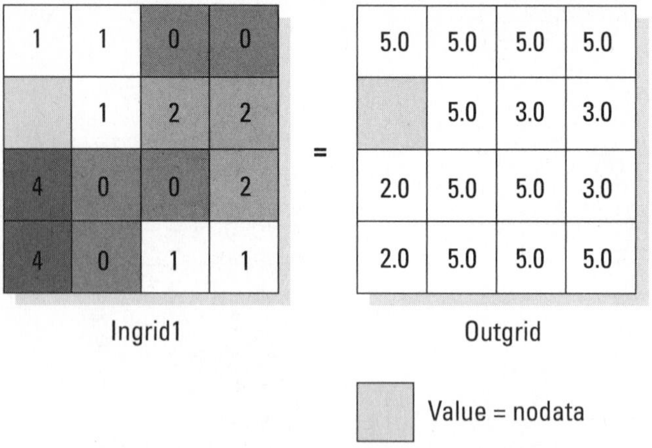

1	1	0	0
	1	2	2
4	0	0	2
4	0	1	1

=

5.0	5.0	5.0	5.0
	5.0	3.0	3.0
2.0	5.0	5.0	3.0
2.0	5.0	5.0	5.0

Ingrid1   Outgrid

Figure 17-11:
A zonal
area.

Value = nodata

Expression: ZONALAREA(INGRID1)

Understanding block functions

Sometimes, you need to resample a set of grid cells to a *coarser* group of grids (bigger grid cells) so that the set matches other raster layers. For example, if a grid has 20-meter grid cells representing 1951 land use, and you want to compare that to Landsat Multi-Spectral Sensor (MSS) data representing 1975 land use in 80-meter pixels (think of these as grid cells derived from satellites), you can generalize the 1951 grid cells so that they match the 1975 pixels. You need to group blocks of grid cells and reclassify them so that they match the 80-meter pixels by using the block function.

The *block function* takes a uniform, non-overlapping block of grid cell values and changes them based on any of the following operators: mean, majority, maximum, minimum, median, minority, range, standard deviation, sum, and variety — the same operators that are available for zonal functions.

The following steps demonstrate how the block function works:

1. **Define how big you want the neighborhood (the block) to be.**

   In the example (shown in Figure 17-12), I define a 3-x-3 block, but your blocks can be any size you want, as long as they stay the same throughout the whole grid.

2. **Define the operator that you want to apply to the block.**

   In this example, I'm using a block maximum operation. So, for each non-overlapping 3-x-3 block, the software looks over all the numbers and searches for the largest number in the block.

3. Create an output grid and assign the values resulting from the operation to each cell in the block.

Now, the difference between focal and block functions becomes really clear. In focal functions (which you can read about in the section "Using focal functions," earlier in this chapter), the software takes that maximum number and assigns it to a single focal cell in the output. In contrast, the block function assigns the maximum number to the entire block: All nine grid cells in each block get the maximum value.

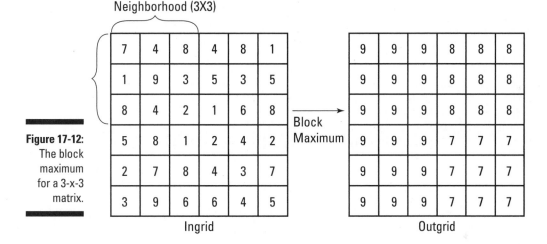

Figure 17-12: The block maximum for a 3-x-3 matrix.

Neighborhood (3X3)

Block Maximum

Ingrid

Outgrid

Using global functions

Global functions are the exact opposite of local functions (as discussed in the section "Using local functions," earlier in this chapter). Instead of seeing the database from a local function's worm's eye view, a global function takes the bird's-eye view — it can see and operate on the entire study area at the same time. Global functions are very powerful and complex operations. Here are some examples:

- **Distance measures:** Such as Euclidean distance, Manhattan distance
- **Surface functions:** Such as finding basins, pour points, and drainage networks
- **Interpolation functions:** Such as linear, nonlinear, trend surface, and exact
- **Hydrology functions:** Such as water accumulation and flow direction

You can also use two other groups of global functions — the groundwater global functions and the multivariate global functions.

The really geeky groundwater global function

The groundwater global functions are some of the most complex functions available in the standard GIS software package. They're almost universally raster based and allow you to not only model the flow of liquids in different *substrates* (permeable material that fluids flow through), but also in different thicknesses of substrate (for example, underground aquifers), and with different amounts of *head gradient* (the amount of pressure based on the slope through which the underground water moves). These functions also allow you to analyze the movement of dissolved solids within these systems, which you might use to model point (occurring at a single source) and non-point source (occurring over an area) pollution.

Many hydrological engineering programs and specialized GIS software packages deal explicitly with subsurface flows and groundwater movement. Many general GIS packages are now incorporating some of these sophisticated operations to satisfy an increasing customer base. I don't want to bog you . . . or me . . . down in the gory detail of these models. If you're a hydrologist, you can find documentation about these functions.

The radical fringe global functions

The multivariate functions of the GIS aren't technically GIS functions at all — they're a collection and implementation of traditional statistical functions, including some very sophisticated techniques familiar mostly to statisticians. I don't go into detail about these functions, but you should know that most professional GIS packages have some of these functions built-in.

Formulating a Model

Whether you use map algebra or any other approach to GIS analysis, you probably use more than one map and more than one analytical technique to answer your questions. These combinations of maps and techniques are generally part of a complex set of *ordered operations* (operations done in a logical sequence) that GIS people call a *model*.

Some call this ordered operation and its output just a model, some a spatial model, some a GIS model, but the term most commonly applied by analysts is a *cartographic model*. It's called a cartographic model partly because C. Dana Tomlin and Joe Berry coined that name years ago, but also because it just makes sense. You're making models that result in maps. Mapmaking is cartography. So, you become, in essence, your own cartographer who makes and combines maps to answer questions. Hence, a cartographic model.

I've talked to literally hundreds of people who specialize in cartographic modeling. Nearly everyone has told me that the key to good cartographic modeling is to formulate the model through the use of flowcharting techniques. They also tell me that you should normally create the flowcharts even before you know what's in your database. Such formulations have helped me decide what should go into the model in the first place.

Making a formulation flowchart

The easiest way to formulate a model is by creating a *formulation flowchart* — a set of graphics and arrows showing steps in the modeling process. A formulation flowchart starts with the final desired product (the spatial information product), breaks it up into sub-models, and finally outlines the actual map elements that you need. Your software may have built-in flowcharting capabilities, you may have your own flowcharting software, or you can even (hold onto your hat here) use a pencil and paper. (Sometimes, this old pencil-and-paper way is the fastest.)

Suppose that you want to put up some solar panels to generate electricity. You call the map that you want to create *solsites,* meaning sites where you should place solar panels. You need to make sure that you fulfill three requirements of your solar-panel placement:

✔ Each site must be at least 2 miles from town.

✔ Each site should be on slopes of at least 30 degrees but less than 55 degrees.

✔ The slopes must face south to get as much sun as possible.

Figure 17-13 shows your formulation flowchart.

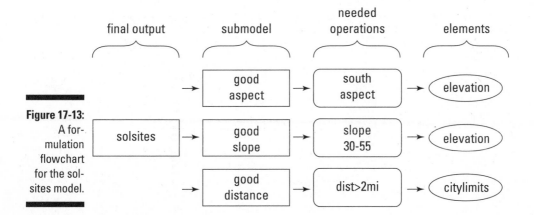

Figure 17-13: A formulation flowchart for the solsites model.

Expanding your sphere of influence

Allocation, which also works with networks (as I talk about in Chapter 15), involves finding all the areas within your *sphere of influence,* or your territory. Say that you and your brother decide that you want to put up small grocery stores at opposite ends of your town. You want to know which part of the town will visit your store and which part will visit your brother's store. In other words, you want to allocate part of the town, and your brother wants to allocate the rest.

You can allocate portions of space to individual facilities in vector systems by using Thiessen polygons, but you have a nice easy-to-understand way of doing it by using raster systems. First, identify the two targets (your grocery stores). The software starts moving out from each of those places to decide which grid cells it allocates to your store and which it allocates to your brother's store. Say that your store has a value of 2, and your brother's store has a value of 1 (see the figure). Depending on your software, it might start its search from your store or from your brother's, but generally, the algorithm that the software uses alternates between the two of them.

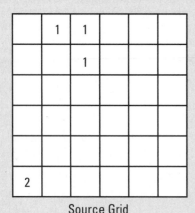

Source Grid   → Allocation →   Allocation Grid

When the software knows the locations of the target cells, it measures distance outward from those locations: 1 for each orthogonal cell and 1.414 for each diagonal cell. It examines each grid cell in the entire database. When it selects each cell away from the target, it compares its distance from each target-cell location. If a grid cell is closer to your store, the software assigns it the same value as your store (2), but if the cell is closer to your brother's store, the software assigns it a value of 1 to correspond to his store. Software differs in how it deals with ties, but most software assigns values based on the order in which it finds the cells. Consult your software documentation to figure out how it assigns ties.

In Figure 17-13, notice how the final spatial information product on the left (solsites) breaks down into submodels that identify searchable aspects of the GIS data:

- ✔ **Good aspect and good slope:** You can derive both of these submodels from manipulations and reclassifications of the same starting map element — topography (essentially, a Digital Elevation Model, or DEM). In this case, you want a south-facing aspect and a slope between 30 and 55 degrees.

- ✔ **Good distance:** To determine an area that fits the good distance sub-model, you measure 2 miles from the outside perimeter of the City polygon on a map layer containing city limits data.

You use the GIS software to combine the sub-models by using a map overlay operation (discussed in Chapter 16) called intersection, which means that it finds only those areas that include all the elements (2 miles from the city, slopes between 30 and 55 degrees, and a south aspect) for your solsites locations.

This basic model has all the elements that you need to make more complex models. You can break models into smaller pieces, and each of the pieces requires one or more map elements to produce it. And you can use the same map element to create multiple derived maps. Cool, huh?

Basing your database on your flowchart

When you finish a formulation flowchart, you know exactly what map elements you need to create the maps you want. Instead of relying on an existing database, you can make all the decisions about what maps, at what scales, and with what classifications you need to get your answer. You're not even restricted by whether the maps exist. You're creating a best-case scenario, and the most effect and most accurate models start with this kind of model-specific data.

Always take the time to develop a formulation flowchart of your cartographic model that includes everything you want to see the model contain without worrying about whether the data are available.

Implementing a Model

Implementing a cartographic model is really just the reverse of formulating it, but with one major difference. You can no longer assume that you have all the best data and all the map elements that you need to do the job. If you assume that you have all the layers at the appropriate scales and classification levels for your solsites model, you simply need to reverse the flowchart so that it moves from basic map element input to the final product (see Figure 17-14).

Some GIS software has built-in flowcharting programs that do more than provide you with a flowchart. They actually populate the portions of the flowchart with data and take that data through its paces. Programs such as the ESRI Model Builder allow you to run the model many times, changing and modifying its formulas and its contents until it performs in exactly the way you want. Such flowcharting programs provide both documentation for how you constructed the model, and an efficient and repeatable way to reproduce it.

No matter what software you use, or whether it contains a modeling tool, your cartographic models probably don't have the exact data, with the precise level of scale, containing exactly the categories available to you. You can deal with inadequate or improper data in three ways:

✔ **Find better data, if you can.** Getting better data is always the best way to deal with inadequate data. In an ideal situation, you'd have the time to collect primary data specific to your needs. Time and resources may limit your ability to collect this kind of data, however. Consult with other GIS users in your area who may possess data that would work for you. If no one has any data that he'll give you for free, you may have to purchase data from a commercial vendor.

✔ **Adapt the model to fit available data.** Although it's not the best solution, you can modify your model to adapt to what data you do have. You may have to leave parts of the model incomplete, be more general in your conclusions, or modify how the model arrives at its conclusions.

✔ **Employ surrogates.** The final solution to data issues in cartographic modeling is to find data that you can substitute for the data that you really need. Such *spatial surrogates* (replacement data) might include housing in place of population, power lines for missing roads, and fire patterns to indicate wind direction and intensity.

You often have to use spatial surrogates based on rather anecdotal evidence. If it's available to you, try to use literature, expert opinion, or even experimental evidence to make the strongest possible linkage between surrogates and the data for which they substitute. You can do so by determining — and supporting with evidence — that one value is a reasonable approximation of the other.

Some of these techniques may require a leap of faith on your part, but remember, no kind of modeling is an exact science. So, I'm providing a disclaimer — I'm not responsible for the improper use of the output! You might want to include a complete explanation of the methods employed to create the output so that the user knows how reliable the model is.

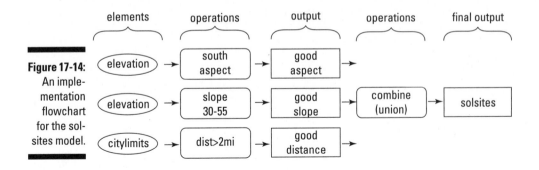

Figure 17-14:
An imple-
mentation
flowchart
for the sol-
sites model.

Testing a Model

No model is perfect. In many cases, you can't get any correct answers — only defensible and indefensible ones. An often-neglected part of cartographic modeling involves testing the output. A true test of modeling often requires a *truth set* (a known correct answer) against which you test the output.

Unfortunately, visiting a location that you modeled probably can't yield answers about whether your model is correct. Imagine, for example, trying to test a model that predicts the population distribution of your city in 2020. Unless you have a time machine, you can't figure out whether the model is a perfect predictor of the distribution.

Over the years, I've discovered four aspects of GIS modeling that you can examine without the need for a time machine. The following sections cover these points in detail:

- ✔ Is the software working correctly (giving you the correct numbers based on the algorithms)?
- ✔ Does the model produce answers that aren't too limiting or too general?
- ✔ Is the outcome logical and consistent with the real world?
- ✔ Is the user satisfied with the product?

Determining whether the software is working correctly

When you need to figure out whether the software algorithms are working correctly, you can most easily perform an analysis on a very small part of the database and check it against a manual computation of the same area.

Assessing whether the model gives adequate results

You need to carefully consider whether the constraints that you put into your model give you adequate results. I've run models that attempt to show the most appropriate place to put a planned development only to have the model tell me that the majority of the unoccupied portion of the city will do nicely. If you get this kind of result, you need to rethink your model and tighten the rules and constraints. I also created a model to show the best location for winter movement of troops, and the output gave me a blank map because I made the constraints too tight. Adjusting model constraints takes finesse; consult the user when you make adjustments so that you don't violate the spirit or purpose of the model.

Gauging whether your model makes sense

Models must make sense. I can't tell you how to determine whether a model makes sense because I'm probably not an expert in your particular field or discipline. Use your own background to assess whether the logic is consistent with how your portion of the real world works. To make this determination, you (and often the user) need to be able to interpret the model. Making your cartographic model understandable, you may need to provide ample documentation, make sure that the model employs easy-to-understand functions and operations, include flowcharts, and even keep the model simple and general enough that users can easily figure out its purpose and function.

Ensuring that your model satisfies the user

Most real-world GIS operations employ consulting firms to do their cartographic modeling (although this fact is changing). The users must be satisfied with the model. They might find a really accurate model worthless because it doesn't provide the answers that they want or expect, they can't read or interpret the output (bad map design), the output is in the wrong format (for example, a travel log versus a map), or the GIS doesn't deliver the model and output on time.

I don't have any easy answers to potential problems of model-user dissatisfaction, but I do have one suggestion: Make sure that you always keep the lines of communication open with the users. If you keep the users involved throughout the process, they're much more likely to accept the final product. A correct model is a useful model only if the user accepts it. Keep the user involved in every phase of your GIS work.

Part V
GIS Output and Application

The 5th Wave   By Rich Tennant

"As you can see, these GIS map overlays show hat use increasing in the uptown area while comb-overs are still dominant downtown."

In this part . . .

A GIS analysis is only as good as its ability to communicate the results to the user. In this part, I show you how you can use GIS to create traditional map output. Or you might opt for some radically different types of maps (called cartograms), which are sometimes even more effective at communicating your message.

But GIS output isn't limited to maps; this part also introduces other types of output, including travel directions, lists, tables, animations, and flythroughs — all of which help boost your understanding of how to present the results of your analyses. Finally, this part offers a guide for how you can efficiently incorporate the GIS technology into your organization and take advantage of its transforming power.

Chapter 18

Producing Cartographic Output

· ·

· ·

*I*n the past, trained cartographers in cartographic production shops made almost all the maps created. Their products were hard copy maps done on parchment, paper, Mylar, or other surfaces, depending on the technology of the day. Today, large production shops are still available in government agencies such as the National Geospatial-Intelligence Agency (NGA) or in businesses (such as Rand McNally) that specialize in maps. Many professional cartographers do both production and cartographic design work for GIS companies, as well.

These large shops will likely be around for some time. They can always find a market for large volumes of well-designed maps, ranging from the general reference and atlas maps to content-specific thematic maps. Although the shops have staying power, their role in producing cartographic output is quickly and permanently changing because of the ready availability and rapidly increasing numbers of GIS software users.

When you work with GIS software, you're the cartographer. Unfortunately, you don't automatically inherit the training and experience that professional cartographers possess because you own the latest and greatest GIS software. GIS software provides you access to the cartographer's toolkit, but many of those tools have specialized uses that require instruction and particular conditions under which you can and should use them.

The cartographic product of your GIS work is important. You can do the most elaborate and impressive GIS models in the world, but when you're done, you need to produce a map that the user accepts and, more importantly, that exactly communicates the analysis.

This chapter shows some basic ideas behind how you can produce good maps. I show you some bad maps and describe how they detract from the quality of your GIS work. The more you work with GIS, the more maps you see. While you see more and more quality maps, you can begin to appreciate all the hard work that goes into them. More importantly, you start to see how the application of some basic principles can lead to effective map output.

Exploring Traditional Maps

Traditional maps are the most common forms of maps and generally the most easily understood by the general public. Reading and interpreting maps is much easier if the scales, symbology, classes, and graphic design elements are all focused on visual communication. (See Chapter 2 for more about how maps show information.) These principles hold true for computer displays, just like they do for maps put on silk, papyrus, or other non-computer surfaces.

Here are the five basic characteristics of good map design (the cardinal rules of mapping):

- ✔ **Make sure that the map meets users' needs.** Always make this characteristic rule number one. If the user can't interpret the map or the map presents the wrong information, all your time and the results of your analysis are wasted.

- ✔ **Make the map easy to use and understand.** Avoid confusion at all costs.

- ✔ **Accuracy is essential to good map design.** As much as possible, avoid data errors, unintended data and graphic distortions, and misinterpretations.

- ✔ **The method of map presentation should relate correctly to the data.** Make sure that you use the right type of map — one with the right symbols, correct classes, scales, projections, and so on.

- ✔ **Allow the user to review and interact with the map during production.** Make sure the user understands what the map is all about, that it meets his or her needs, and that it communicates the information correctly.

Mapping qualitative data

Sometimes, when you create a map, you map *qualitative* (descriptive or nominal) data, which generally requires that you pick unique symbols for each category.

Many software programs set their default so that when you open your map of a certain set of polygons, the map contains the borders for each polygon, but the polygons themselves are all the same color. You may find this layout confusing the first time you see it, but don't worry. All your information is there. The software defaults to this layout so that you can determine the colors and patterns you want to use for your map, as shown in Figure 18-1.

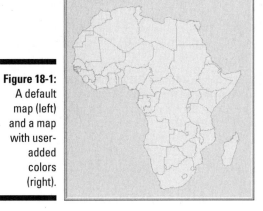 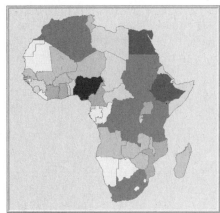

Figure 18-1:
A default map (left) and a map with user-added colors (right).

The GIS groups the categories by attribute value, such as land-use type or road type. Using a separate pattern for each category allows the reader to instantly pick out one from another. Two general types of attributes assign unique colors or patterns to category values:

- ✔ Attributes that describe the names, types, or qualitative condition of the features
- ✔ Attributes that identify specific objects, such as the names for countries, states, or colors

Most GIS software allows you to use a set of predesigned color or shading schemes, or to create and employ your own. When working in black and white (for publications that don't use color, for example), you can use the shading patterns rather than colors.

Mapping quantitative data

Most of the output from GIS is based on numbers, statistics, and other quantitative data that describe something about the features you're mapping.

Most non-cartographers don't really know how to decide what specific data to map or how to display them. Here are three general rules to help out:

- ✔ **If you want to see the raw data and relative comparisons, map the counts or amounts themselves.** Be careful, however, because amounts might be misleading. Imagine looking at numbers of newspaper subscribers in New York City — millions of people — versus Bismarck, North Dakota, with less than 100,000 people. The results of such a map merely reflect total population rather than the number of newspaper readers per capita.

✔ **If you want to control for area or number of features (called *normalizing your data*), use ratios.** Ratios require you to compare two values. For example, you could compare the number of newspaper subscriptions per 1,000 people living in the subscription area or population per square mile.

✔ **If you want to compare the relative values from place to place or feature to feature, you want to map those values as ranks.** You might rank newspaper subscription ratios per 1,000 people for each city. So, you replace the actual numbers with the ranks.

Buying an output device

Even though computers can display maps on-screen, hardcopy still has its charms. People often print out digital text documents, even though they don't have to, partly because people are usually more comfortable when they can actually touch a map because they're usually more familiar with maps as documents.

GIS software and hardware can produce hard copy maps. You can find many devices that are designed just for these tasks. Some devices print only text, and others print graphics, as well. Keep these tips in mind when you're ready to go from your GIS to hard copy output:

✔ **Choose an output device that's right for your needs.** Large-format plotters and raster printers can produce large maps, but they require special paper, inks, and supplies that affect your budget.

✔ **Choose output devices based on volume.** If you plan to make only a few production maps, you need only one set of devices, but if you plan to create a production facility, you need to consider much higher volume production equipment.

✔ **Consider the ink.** Most inks fade with time, but some are more prone to fading than others. Ask your vendor about the impact of light on the ink. Also, keep in mind that the colors of the ink are *not* the same as what you see on your monitor. Many GIS users are disappointed by the loss of color intensity, lack of contrast, or even the different appearance of the colors.

✔ **Keep supplies handy.** You never know when a big job will come down the road and you'll need supplies. You need time to get your hands on some kinds of supplies. Find a good, reliable vendor. Make sure the vendor knows your needs and the demand you place on the equipment and supplies.

✔ **Have a backup plan for output.** Having supplies such as toner drums, ink, pens, and paper doesn't help if your printer stops working. Have additional output devices as backups or reach an agreement with another organization that can help you in a pinch.

✔ **Check compatibility.** Output devices have their own graphics languages. Make sure that your GIS software and your output device can talk to each other. You normally don't have to worry about a problem if you use professional GIS software that either comes bundled with compatible hardware or is adaptable to changes in hardware.

Creating classes

An important aspect of deciding what to map deals with whether you want to map each individual value, or group (classify) the values and then map those groups. If you have only a few items to map (normally, less than ten), you can display each one with a separate symbol or color. If you have more than ten items (which you usually do), you need to group those items. Otherwise, the reader can't visually separate the many shades and colors needed to display them.

For example, if you tried to map each of the 3,000 or so counties of the continental United States with a unique color or pattern for each county, the map reader would have to try to distinguish between these 3,000 colors and patterns. For the map to be readable, you'd have to create groups (classes) that reflect real differences in value and enable the readers to see the patterns you want them to see.

You can create classes in many ways and use each class in a variety of ways. You want to put data into classes that allow similar values to be in the same class — sort of like how teachers separate out student exams into letter grades. You need to make two decisions:

- ✔ How many classes you need
- ✔ How to divide the classes into appropriate groups

You typically find seven general ways to classify data available in GIS software. Different software uses different names for these classifications, but here are general descriptions:

- ✔ **Equal interval:** Equally divides the range of values to be classified, for example, 0-100, 101-200, 201-300, and so on. This method defines the number of classes, as well as the class intervals. This classification works best when you have data ranges that are fairly commonly understood and well known, such as percentages or temperatures.

- ✔ **Defined interval:** This method allows you to define the specific ranges of data and lets the software decide how many classes you need. Suppose you tell the software that you want it to divide the percent of people who voted in the last election into intervals based on 5-year age groups. But you've already restricted the dataset to only those people between the ages of 30 and 59. In this case, your software will create six classes reflecting your choices: age groups 30-34, 35-39, 40-44, 45-49, 50-54, and 55-59.

✔ **Quantiles:** This method counts features so that each class has the same number of features (for example, the same number of counties). If you know that your data are distributed in a linear fashion, you can use this approach effectively. If wildly different data occur together or very similar data occur in different classes, you can increase the number of classes you use to better represent the data classes using quantiles.

✔ **Natural breaks:** I'm rather fond of this approach because I can see the visual changes in the categories and put in the class breaks where they occur naturally. Classes created this way are more likely to be representative of how data should logically be classed. The visual changes occur where the data have large jumps or dips so they are very easy to pick out.

✔ **Geometrical intervals:** Many datasets change in non-linear fashion, for example, logarithmically. This approach, designed for continuous data, attempts to put the same number of data occurrences in each class regardless of what their values are.

✔ **Standard deviations:** If you're sure that your data are distributed in a bell shape (that is, normally distributed), you might want to use this method because it's based on sound statistical theory. This classification method is designed to use the standard deviations above and below the mean, which identifies the category value based on that standard distance from the mean. ESRI recommends using two colors when you employ this method — one color for those values above the mean and one for those below the mean.

✔ **Area-based:** Only some software offers this approach. It allows you to use a variety of the other methods, but it divides the classes based not on the distribution of the data, but rather on the distribution of the amount of area for each category. This approach gives you quite a bit of control over the visual appearance of the map and adjusts for graphical artifacts such as disproportionately large polygons that skew the map's appearance.

Using map elements

Map elements are the graphic devices that cartographers include on a map. Here's a list of the standard map elements (see Figure 18-2):

✔ **Title:** Gives the reader a context for understanding the map.

✔ **Legend:** Explains what the content and classes mean.

✔ **Scale:** Helps the reader know how big the area is and how much generalization to expect.

✔ **Credits:** Hey, here's where you can take credit for your hard mapmaking work. Nothing wrong with that!

✔ **Mapped areas:** The place where you actually put the geography.

✔ **Graticules:** A grid of fine lines used to determine scale and position of mapped objects. Maps don't always need graticules, but they often make a map look really professional.

✔ **Borders:** Sometimes called *neat lines* because they surround the map area and make the map look orderly and contained.

✔ **Symbols:** You can't really put the mapped areas on the map without using symbols that represent the real geography.

✔ **Place names:** The names of the locations that are mapped so the reader can find them on the map.

Keep in mind that map standards are always changing, so these elements may be different from what you've seen in the past — and they may change in the future.

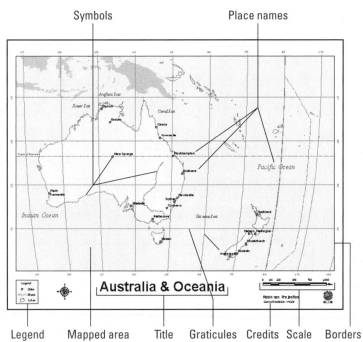

Figure 18-2:
Anatomy
of a map.

Legend   Mapped area   Title   Graticules   Credits   Scale   Borders

Purchasing a monitor

Before you ever put ink on paper, your GIS displays the maps on your computer monitor. Monitors come in a wide variety of shapes, sizes, resolutions, and contrast levels. Some monitors are CRT technology (based on old-style television tubes), some are LCD (liquid crystal display), and others are plasma-based. Some monitors can handle millions of colors, and others are more limited.

Your visualization needs, budget, and even lighting conditions can affect your decision about what type (or types) of monitor(s) you need. Some systems employ two monitors so that they can separate graphics and text. You don't need a high-quality, high-resolution monitor for the text side of the system, but you might for the graphics side. Some monitors look washed out if you view them at an angle or surround them with too much light.

GIS vendors often bundle the hardware to their software. You can often get excellent compatibility between the software and the hardware in these bundles because the vendors are familiar with the capabilities of their software relative to the hardware. Even if you don't purchase a package that includes both your GIS hardware and software, you can ask your software supplier which monitor systems it recommends for your application. Suppliers normally have many different types of users, and the users provide very good feedback on what works and what doesn't. Here are some basic aspects of graphics monitors for you to ask about:

✔ Resolution

✔ Size

✔ Refresh rate

✔ Material (whether it's CRT, LCD, or plasma)

✔ Number of colors

✔ Number of monitors (usually one versus two)

✔ Reliability

✔ Compatibility with the software

✔ Vendor responsiveness and warranties

When you ask your GIS vendor about these specifications, be sure that you know your needs. In particular, be able to explain your lighting, space configuration, user requirements, and ergonomics. The vendor may want to know how many stations you need and whether any of your technicians have visual or other impairments that affect seating and viewing. If one of your technicians, for example, is in a wheelchair, his or her monitor needs may be quite different than those who will sit in an office chair to view the monitor, depending on the desk configuration. You may also have users who are colorblind and therefore don't need a massive color palette (as long as the monitor can display plenty of gray shades). In some cases, you might want to have a very large monitor or a projection system so that you can share the results or the preliminary analyses with your clients comfortably.

Factoring in graphic map design

A map offers more than just a way to store and retrieve data. Its primary functions are analysis and display. Selecting the correct data, scales, projections, classes, and map elements helps you make sure that you're creating a readable map. But you need to remember that the map is also a graphic device — and some graphics are nicer than others. This aspect of cartography, called *graphic map design,* borrows heavily from traditional art and graphic design principles, as well. Here are a few questions to ask yourself while you look over your finished product:

✔ **Does the map look unbalanced (not visually leaning to one side)?** If it does, you need to move the map elements around. Use the text, titles, and legend to help you balance the map graphically.

✔ **Can you read the lettering?** You need to know how close the readers are going to be to your map. Put yourself in their shoes and increase the lettering as needed.

✔ **Do the features stand out?** Contrast is the basis of seeing. Putting pink on yellow might work for a designer hat, but this color combination doesn't help a map-reader see the objects you want him to see.

✔ **Does the reader know what to focus on?** Use a technique called *figure ground* where you make sure the background is rather vanilla and the *figure* (what you want the reader to see) stands out. You may also want to use *visual hierarchy* — using line thicknesses to help distinguish the importance of geographic features such as nations, states, and counties.

✔ **Do you feel dizzy looking at your map?** This question may sound funny, but it happens. Poor colors, colliding line patterns, and spiraling shapes make readers uncomfortable — which means they don't want to read your map, and you've failed as a map designer.

Over time, you become sensitized to what works and what doesn't work in graphic map design. When you look at art, you know what you like and what you don't like. The same is true of maps. After you become familiar with the terminology of art, you can explain why you like or don't like art. In the same way, when you become familiar with cartographic design language, you can do the same with maps.

Look at a lot of maps. If you want to be a better mapmaker, you need to see what works and what doesn't work. If you go to GIS conferences, you can see tons of maps, both good and bad.

Understanding Cartograms

For most GIS cartographic output, the user needs the traditional map. Most users are most comfortable with the typical thematic map output if you keep them apprised while the output progresses. In almost all situations, the usual types of maps make their point quite nicely. Sometimes, however, you might want to make a point that the usual display just doesn't convey with the impact you want. When you really want to elicit a reaction from the map reader, you can use a really cool type of map called a cartogram. A *cartogram* is a map that exaggerates the size of geographic space based on the descriptive attributes, instead of presenting the space in its accurately scaled size and shape. You use a cartogram to communicate a specific aspect of the mapped information more effectively.

In traditional maps, the sizes of the land areas and the lengths of line features portray (as closely as possible) the proportional relationship in their scaled size. This proportional look makes sense, of course, when the features and their relative sizes are familiar to the audience. When you distort these area sizes or change the lengths of features, the reader becomes uncomfortable. This discomfort may be useful to elicit the response you want, or it may detract from that response if the viewer is unfamiliar with the area portrayed on the map. You must be aware of the target audience's map reading skills and understanding of the portrayed geography before attempting to use cartograms.

Although you don't want to go out of your way to make your reader uncomfortable with an out-of-proportion map, sometimes the shock of such distortions has the impact on the viewer that you want. Impressionistic artists use this technique in their work to elicit a response rather than to represent reality as accurately as possible. Cartographers use distortion to draw attention to a particular result. These distortions are performed on two types of map elements: areas and distances.

Attracting attention with area cartograms

Area cartograms, often called *value-by-area maps,* distort the sizes of the polygons based on the attribute quantity, rather than its actual physical size. Among the most common examples of a value-by-area map is the analysis of U.S. presidential election results, like the one shown in Figure 18-3. In this figure, the top map is a traditional *choropleth* map (a map presenting data on an area-by-area basis) that shows the actual scaled size of the United States. The darker states (the West Coast, East Coast, and areas near the Great Lakes) voted for the Democratic candidate, and the remainder (what appears to be the vast majority of the country) voted for the Republican candidate. But when you distort the areas based on the states' *electoral votes* (the number of votes each state gets based on its population), the election results seem much closer than in the original map.

Use value-by-area maps to force the reader to focus on the value of the attribute represented by each area (in Figure 18-3, the electoral votes for each candidate), rather than on the size that the map features take when accurately scaled.

Figure 18-3 shows that all the states (polygons) are connected. This type of area cartogram is called a *contiguous cartogram*. You can also make an area cartogram by exploding the polygons so that they're relatively close to their proper locations but no longer connected. This type of area cartogram is called a *non-contiguous cartogram* (Figure 18-4).

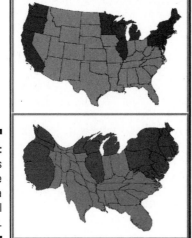

Figure 18-3:
Two maps showing the results of a presidential election.

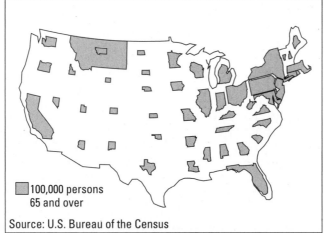

Figure 18-4:
A non-contiguous area cartogram maintains the features' shapes but not their exact locations.

100,000 persons 65 and over

Source: U.S. Bureau of the Census

Contiguous and non-contiguous area cartograms don't have any real advantages or disadvantages. Contiguous area cartograms distort shapes, and non-contiguous area cartograms distort relative positions. When deciding which cartogram style to use, consider your potential map readers. The less cartogram experience the readers have, the more they rely on shape, so the non-contiguous style might make sense to them. However, if the shapes are unfamiliar to your readers, the contiguous, shape-distorted style may be as good a choice because the reader is not relying on shape to interpret the map.

Distorting distance with linear cartograms

Linear cartograms distort distances to make their point. Perhaps the most famous linear cartogram, and arguably the first published in the geographical literature (research papers written by geographers), is called the central-point cartogram (see Figure 18-5). The *central-point cartogram* illustrates the time it takes to travel from a central point to every other place in a city. When travel time lines are displayed with traditional mapping techniques, the time lines are distorted based on the impact of the road network on travel. In Figure 18-5, the travel time lines are concentric, so the map of the city is distorted so that the time travel lines can remain circles. You can display a network analysis this way, rather than with a typical map. (Chapter 15 covers network analysis in detail.)

Mapping sequence with routed line cartograms

Perhaps the most used, but often unrecognized, linear cartogram is a *routed line cartogram*. Busses and mass transit trains display this kind of cartogram to illustrate the stops in their proper sequence. When you travel on mass transit, you don't really need to know the distance you travel. You do need to know when your stop occurs within a linear sequence of stops, though. To create this type of map, you ignore the distance between stops and include only the sequence of stops. Figure 18-6 shows an example from a train system.

Use routed line cartograms if the sequence of places is more important than their distances from each other.

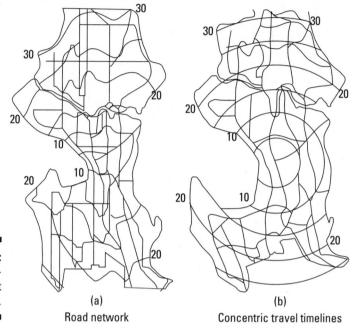

Figure 18-5:
A central-
point
cartogram.

(a)
Road network

(b)
Concentric travel timelines

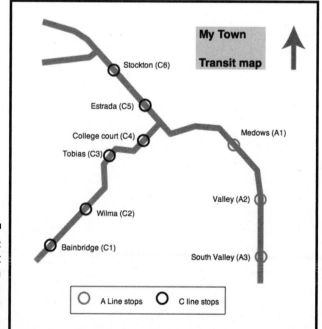

Figure 18-6:
This transit
line system
map is a
routed line
cartogram.

Chapter 19

Generating Non-Cartographic Output

M aps are the traditional output from GIS, but not everyone is well-versed in reading maps — especially thematic maps (which I talk about in Chapter 2), the typical type of maps resulting from GIS data retrieval or analysis. I've been to several conferences where people got into lengthy, often heated, discussions about exactly what a particular map was trying to communicate. Such disagreement often happens because of poor cartography, but it can also happen because of the map reader's lack of familiarity with map symbols, classification strategies, and legends.

The ultimate goal of GIS is to communicate results. Many people are familiar with lists, tables, statistics, travel directions, and other non-map-like forms of communication. If the intended recipients of your GIS output are more comfortable with these alternatives to maps, then why not provide them with these forms of communication, rather than — or, at least, in addition to — cartographic output?

GIS has many ways to communicate results that are not purely cartographic. This chapter helps you with a few of these methods so that you're ready to deliver your GIS output to meet the needs of diverse audiences.

Looking for Routings and Travel Directions

GIS provides many opportunities to find shortest routes, fastest routes, and even the most scenic routes along a road network. (Chapter 15 discusses such routes, which you can identify in networks.) Such analyses often give you maps that highlight the route. In some cases, a non-graphical list of turn-by-turn directions accompanies the map, as shown in Figure 19-1. Travelers charged with navigation duties often prefer this type of output because they can refer to it and reference the information piece by piece.

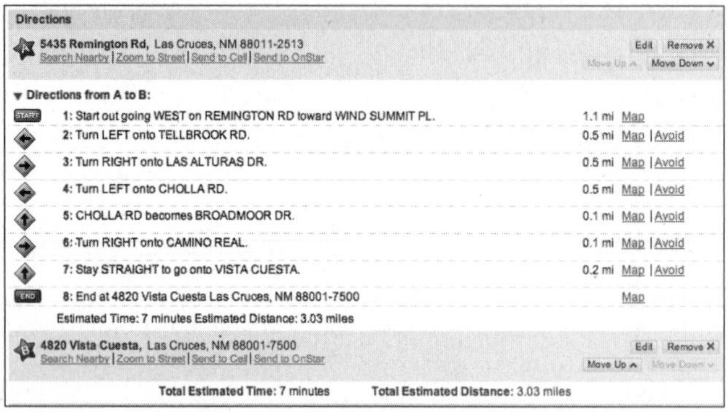

Figure 19-1: Turn-by-turn output from MapQuest.

Travel directions from a service such as MapQuest (www.mapquest.com) are normally based on shortest-path distance, but GIS software (and even online mapping systems) allows you to request directions based on time, rather than distance. In some instances, the software even allows you to select which specific streets you want to avoid. Road and highway icons that resemble their respective road signs add a bit of color to the output and can also help the readers find the individual streets.

Getting Customer Lists and Statistical Data

The business GIS user often wants information about customers or potential customers. One typical application uses GIS to search business databases for names, addresses, purchases, e-mail addresses, and a host of other information that a business uses for target advertising. In some situations, every time you scan a purchase and use a credit card, the store records who you are and what you purchased. By comparing that record to *demographic data* (selected characteristics of the general population), businesses can actually target different parts of a city for specific marketing campaigns.

Although maps can provide a general description of customer type and location, the busy business professional often wants a report, rather than a map. In some cases, the report includes maps as part of the document. GIS produces two general types of reports that are suited to business uses:

✔ **Tabular reports:** Have a spreadsheet-like layout

✔ **Columnar reports:** Have a newspaper column layout

Whether you choose the tabular- or columnar-report format, you can sort by any data field that the report contains. For example, you can list cities alphabetically, by population, market share, or total volume of business. Another useful tool in the GIS reporting toolkit generates summary statistics, such as sum, mean, range, standard deviation, and minimum and maximum for any data field.

Typically, you can more easily understand summary statistics when they're presented graphically through bar charts, histograms, line graphs, pie charts, scatterplots, and other forms of non-cartographic graphic output. Most GIS software gives you the option of including both tables and graphs directly on the map itself. Figure 19-2 shows a graphic presentation of summary statistics.

Figure 19-2:
A graph showing summary statistics.

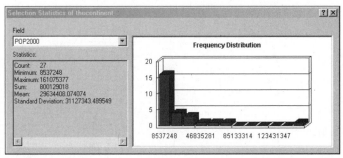

GIS software often presents statistical and tabular data as graphs because graphs provide a visual, easy-to-understand means of communication. Used alone, graphs are handy, but GIS analysts often use them to complement a map because the reader can gather information faster from the graph than from only the map itself. Most GIS software allows you to not only produce maps of GIS data, but also non-spatial data associated with the map. When you base this non-spatial data on the map attributes, the GIS often links that data to the map so that when the data changes, so does the graph. You can even use the same color schemes to make comparisons more natural.

In addition to generating maps, a GIS can present virtually any of its data as statistical tables or graphical charts and diagrams. You have a staggering volume of potential output. Here's a list of some practical and frequently used GIS output that works well in non-cartographic formats:

- ✔ Market share by region or quarter
- ✔ Number of customers in a particular city, county, state, or neighborhood
- ✔ Lightning strikes by state
- ✔ Customer or potential customer profiles by region
- ✔ Crop yields and changes in crop yield by year
- ✔ Lists of households not currently subscribing to a newspaper
- ✔ Lists of property tax valuation for counties

You can more easily read and interpret maps if you include a linked table or graph either on or near the map itself (as shown in Figure 19-3). If you use matching colors to show the correspondence between the map symbols and the tabular or graphed data, you can also improve users' understanding.

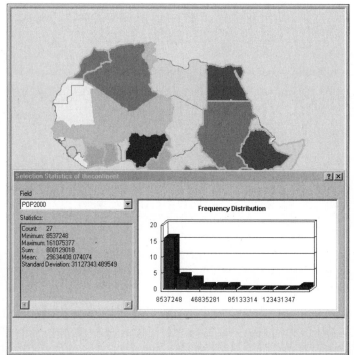

Figure 19-3: Combining a map with summary statistics.

Producing Alarms and Signals (Audio and Video)

Sometimes, GIS produces output that it can't put in a map or graph. Some home security systems, many E911 systems, and a growing number of marketing and government bodies now use automated electronic and signaling systems that, at least in part, employ GIS technologies. Here are some examples:

- ✔ The output from analysis of registered voters could result in automated telephone calls asking for your vote.

- ✔ Fire and emergency services may receive signals from dispatchers that include both an audible alarm and even a set of routing directions in the vehicle's GPS unit.

- ✔ Some military hardware systems combine GPS, GIS, and remote sensing technologies to direct automated aircraft to particular targets.

- ✔ In precision agriculture solutions, certain combinations of input trigger valves to open and close on machinery that releases specific amounts of fertilizer or pesticide.

- ✔ Real-time information is sent between GIS operators and forest-fire fighters.

- ✔ Automobile onboard navigation systems often have voice command and even updated traffic information as part of their output.

Benefiting from Virtual Output

With newer, faster computers and better graphics output devices, GIS has a new set of possibilities for presenting GIS data and analytical results. The following sections show you two major types — animations and flythroughs — both of which involve the development of short movie clips. These types of video output give GIS users a totally different perspective on the data.

Animating your maps

GIS *animations* are a form of geographic visualization that shows dynamic changes to a single map layer or even multiple map layers simultaneously to allow visual comparisons. You most commonly use GIS animations to depict geographic movements (for example, tracking the movement of wildlife) and temporal changes (such as changes in land cover patterns over many years).

Animations show changes — which might take days or weeks to occur — at a fast pace so that you can see trends you might otherwise miss. Alternatively, if you have a lot of data about changes that occur very quickly, you can use animation to present those changes slowly and, again, identify changes you might not otherwise catch. Some circumstances might warrant moving through an animation backward and forward so that you can focus on specific events in the sequence.

Here are several possible uses of animations in GIS:

- ✔ Tracking hurricanes and other weather-related phenomena

- ✔ Keeping tabs on the movements of delivery trucks that have built-in GPS units

- ✔ Watching changes in climatic data, such as temperature, pressure, and precipitation, over months or even years to observe climate change

- ✔ Viewing airline flights to detect possible locations for high traffic concentrations and potential accidents

- ✔ Tracking radio-collared wildlife species to identify habitat preferences

- ✔ Examining land-use change through time to determine major trends

- ✔ Observing how crime hot spots change throughout the day, during the week, or even at different seasons

- ✔ Watching changes in market share of your own and competing businesses to identify where new businesses are impacting your sales

Use GIS data presented as animations when you want to focus on pattern changes or feature movements through time.

Getting the most from flythroughs

The *flythrough* is a form of animation that happens in three dimensions. Just like the term says, the flythrough allows you to visually fly through a three-dimensional version of your map. You can find a common example of flythroughs in software such as Google Earth, but most professional GIS packages also possess this capability and give you total control over how the flight moves. You control the flight path or track, perspectives, and camera locations of the flythroughs in two ways:

✔ **By setting up a sequential path:** You select a number of key locations that you want to visit in the sequence that you want to view them to create the general flythrough path.

✔ **By recording your movements during a flight simulation:** You control the speed of the flythrough based on the speed at which you move during the recording process.

Whether you set up a path or use the simulator approach, the software records a sufficient number of frames per second so that when the flythrough is played back, it moves smoothly.

Flythrough is useful for the following GIS applications:

✔ **Terrain evaluations:** To detect landscape features or compare terrain texture (ruggedness or smoothness)

✔ **Urban impact assessment:** To show the effect of urban structures on wind

✔ **Political and cultural studies:** To illustrate the often radically different patterns that occur as you cross political boundaries worldwide

✔ **Damage estimation:** In which flying through an entire valley allows you to observe flood conditions

✔ **Ecological research:** Making visual comparisons of elevation levels to changes in the size and composition of vegetation

I used this technique to show how datum errors can result in strange mismatches of aerial features on topography. As shown in a snapshot of a flythrough around Tortugas Mountain in Las Cruces, New Mexico (see Figure 19-4), the letter A was turned upside down because the Digital Elevation Model (DEM) and the aerial photography used different datums. (I talk about datums in Chapter 2.)

Vendors usually save flythrough animations in some proprietary format, but in most cases, you can also export flythroughs as Audio Video Interleave (`.avi`) or Quicktime (`.mov`) files that you can play back by using Windows Media Player or QuickTime.

Use flythroughs when you need multiple views of a topographic feature or when you want to see the relationships that might exist between elevation and other features on the surface.

West side of "A" Mountain

East side of "A" Mountain (DEM)
Notice A is upside down

Figure 19-4:
Flythroughs
can point
out data
errors.

East side of "A" Mountain
(Corrected DEM)
Notice no A present

Chapter 20

GIS in Organizations

In This Chapter

▶ Understanding how organizational structures affect GIS integration

▶ Knowing the types of organizations that benefit most from GIS

▶ Incorporating GIS in your organization

GIS is much more than just software and hardware. You need data, training, space, personnel, funding, technical support, and many other elements that work together to make your GIS function properly. Your organization must undergo fundamental changes in the way it functions, both in its internal and external interactions. You can make these changes extremely positive, if you incorporate GIS effectively. The guidelines in this chapter help you ensure that when your organization adopts GIS, the benefit outweighs the cost.

Understanding How Your Organization's Interactions Change

GIS fundamentally changes the way an organization works. Not only does your organization change, but so do interactions within the organization. You interact with *internal players,* meaning your organization's GIS operators, managers, and other GIS personnel. But you and they also must obtain data, supply output, purchase equipment, get training, and participate in many other tasks that involve the larger GIS community, called *external players.*

Internal players include the following general groups:

✔ **System users:** Use GIS to solve problems, provide solutions, and supply decision support for the organization. Their skills generally include spatial problem solving (for example, performing the actual analysis on the data) and model building (such as systematically applying large numbers of analytical techniques to solve larger problems).

✔ **System operators:** Know the inner workings of the software, can support the system users through troubleshooting and workarounds, and have specialized knowledge about the GIS. They usually provide security and training, as well. These people are highly knowledgeable about the capabilities of the GIS and its place in the organization, have advanced technical training, and work with the GIS every day.

✔ **System sponsor:** Provides the financial support for the entire system. The system sponsor isn't a person — it's the organization itself, or a subdivision or unit within the organization. This group provides the funding for data, hardware, software, space, utilities, training, salaries, and anything else related to the GIS. Without the system sponsor, the GIS operations don't exist. The system sponsor also makes sure that the system remains necessary to the organization, thereby guaranteeing long-term viability and preventing losses from abandoned GIS investment.

External players include the following general groups:

✔ **GIS supplier:** The GIS vendor that supplies the software (and sometimes hardware), technical support, and training.

✔ **Data supplier:** Two types of data suppliers exist, public and private:

• **Public sources:** Often government organizations or public GIS data clearinghouses, including universities that provide data on an as-is basis (often free or inexpensively, as a cost recovery)

• **Private organizations:** Typically charge for data but often provide value-added data that's targeted to your needs

✔ **Applications developer:** One quick way to address short-term, project-specific needs is to hire outside professionals, called *applications developers,* to (guess what?) develop an application that has just the right data, analytical models, and user interfaces.

✔ **GIS systems analyst:** Systems analysts look at the overall operation of your GIS to ensure its initial and long-term successful integration into the organization. Many systems analysts, sometimes called *system designers,* are employees of the GIS supplier or software company that you purchase from. After all, who knows the capabilities of the system and how it interfaces any better?

Categorizing the Types of Organizations That Use GIS

GIS may be integral to the operation of an organization (enterprise); provide data storage, archiving, and retrieval; or act simply as an occasional tool for specific analyses on an individual department basis. These levels of organizational integration are often not static. Instead, they change when an organization recognizes the potential usefulness of the GIS for other branches of that organization.

Some people try to categorize the types of organizations that use GIS based on their level of integration and the ways that they use the GIS. I find this approach a bit difficult to deal with because both organizations and the systems they incorporate grow.

Growth is a fundamental characteristic of any successful organization or system, and this growth changes the way GIS works in an organization through time. To design a GIS based on its initial incorporation for limited use — rather than for its likely eventual adoption on a larger scale — forces a constant re-evaluation of the design process itself, in addition to the organizational GIS design. So, rather than take this somewhat untenable approach, I show you what I call the basic three types of GIS organizations in the following sections.

Private/commercial

Many GIS designers speak of the spatial information product (SIP) — the planned output from GIS analysis. In most private and commercial business operations, the SIP is a product that provides income to the company. Some commercial companies produce output (SIPs) that they make available for purchase, and others produce output for some other organization that has limited GIS capabilities. In the latter case, the output still provides income to the commercial firm.

Here are some example private and commercial company types that use GIS and the SIPs they might produce:

- ✔ **Mapping companies** use GIS to produce maps for sale to the general public.

- ✔ **Consulting firms** produce viable GIS models for organizations that operate GIS but lack the necessary personnel resources to create their own models.

- ✔ **GIS data providers** develop clean, user-based, value-added datasets for other organizations so that those organizations can spend less time converting data and more time analyzing those data.

- ✔ **Economic placement companies** develop location and allocation plans for businesses so that those businesses know the most economically feasible place to put their next store.

- ✔ **Real estate companies** link their GIS software to the Multiple Listing Service (MLS), providing its agents with the means to locate properties, provide virtual tours, and link properties on a map.

- ✔ **Vineyards** use GIS to schedule plantings, decide on watering and applications of supplemental fertilizer, manage harvests, and control workflow during different times of year to guarantee the highest yields.

The applications in the preceding list have one thing in common — the development of a tool (a SIP) that improves the income stream of the organization. Such organizations are structured around that goal. GIS designs must complement that same goal — otherwise, the company won't find the GIS very useful and won't implement it.

Government

A modified version of the spatial information product idea works in government organizations, accommodating the output needs of such organizations. Governments and government-related organizations are generally in the business of controlling and protecting large groups of people through legislation and mandates. And so, the GIS solutions that these organizations use provide information about people, as well as the factors that affect them (for example, natural and man-made resources, urban and infrastructure development, weather and environmental conditions, and so on).

Here are some examples of governmental bodies that use GIS:

- ✔ **Management agencies:** Control and protect the many aspects of the environment, including land use, agriculture, minerals, forests, ocean resources, parks, recreation areas, scenic natural areas, and many others.

- ✔ **Law enforcement and corrections agencies:** Identify, target, and prevent crime, as well as control and manage people who are incarcerated or on parole.

- ✔ **Medical agencies:** Monitor disease, isolate causal factors for outbreaks, allocate medical personnel to infected areas, and even decide where to place medical facilities.

- ✔ **Meteorological and climate centers:** Predict short-term weather phenomena, protect people and property from short-term severe storm events, and determine the long-term impacts of climate change.

- ✔ **Emergency management agencies:** Serve the needs of victims of natural and man-made disasters. Examples of such disasters include floods, hurricanes, tornados, earthquakes, chemical spills, riots, and conflict.

- ✔ **Military and military support organizations:** Protect the nation against armed aggression through intelligence, surveillance, military planning, troop and hardware deployment, and logistics.

All the organizations in the preceding list are mission oriented. They typically have mission statements that clearly and concisely outline their goals, objectives, and mandates. Any GIS design for such organizations must target the fulfillment of those mandates. Otherwise, the organization won't adopt, or successfully implement the GIS.

Non-profit/educational

Conservation, relief, philanthropic, research, and educational organizations and foundations aren't in operation to make money, nor do they have specific government mandates. They often decide their objectives based on their own mission statements, but those mission statements are highly variable because of the wide variety of tasks, and change very quickly as a response to changing demands and circumstances. Some organizations, such as research organizations, don't actually know what their final output might be. Instead, they need the flexibility to perform as wide a range of analyses as possible.

Here are some examples to demonstrate the wildly different non-profit organizations and the ways they might use GIS:

- ✔ **A relief organization** needs to manage the collection, packaging, and dissemination of food and medical supplies to a third world nation that has few roads and little equipment.

- ✔ **A conservation group** is trying to find land that has a high level of wildlife habitat diversity and owners who are willing to negotiate conservation easements or the outright sale of the land.

- ✔ **A university department** wants to develop a combination teaching and research laboratory that can support equipment, software licensing, datasets, and continued software training for extended periods of time.

- ✔ **a philanthropic organization** picks people for funding based on their contribution to research that results in improvements to the human condition. The organization needs to be able to identify these individuals and wants to spread such awards around geographically.

- ✔ **A college admissions department** wants to target students from across the nation whose entrance exam scores and scholastic potential are in the top 5 percent of their graduating classes.

You can design a GIS for organizations that have variable needs (such as the examples in the preceding list), but you need to exercise great care. The spatial information product (for example, the list with names and addresses of targeted students) might be the primary need for the GIS. But in some cases, simply having the system in place for a long time might satisfy a major need that employees have for flexible access and opportunities for experimentation and modeling. Some organizations also have the added complexity of rapid turnover in personnel. So, their training needs may be more important than the hardware or software needs. In some cases, these kinds of organizations use GIS only periodically, so they have minimal operational needs. In these situations, the organizations may prefer to outsource their GIS operations rather than incurring the expense of operating an in-house system.

Designing and Introducing a GIS for Your Organization

Each organization is unique. Each has its own culture, organizational structure, history and traditions, needs and mandates, and vision. These differences have a profound impact on whether the organization can effectively use GIS, or any technology, and whether it can easily integrate that technology into the existing operations. So, you need a thorough knowledge of the structure of an organization — including how decisions flow from place to place, who has what job responsibilities, and how general operations work — to be able to incorporate GIS effectively into the organization.

You can really begin to understand an organization's structure by creating, or having a consultant create, a detailed organizational diagram. Such a diagram helps define where the GIS fits within the organization and shows who answers to whom. GIS is about sharing data resources, and an organization's structure can help you see how, where, and to what extent data sharing takes place.

Organizations don't necessarily process and share data in the same fashion, but instead, have differing resources, approaches, and methods, as follows:

- ✔ Some organizations' systems are designed to widely acquire data and information — but to disseminate it to only specific individuals. GIS output must follow the data path that exists in your organization.

- ✔ The technology of GIS requires a fundamental understanding of both geographic principles and technology. GIS training becomes very important if you find that your organization is composed largely of people who have avoided both technology and geography.

- ✔ Some organizations have very rigid hierarchical structures with equally rigid paths for operations and information flow; others are loose and flexible, so responsibilities and information flow freely among employees. The GIS operations should fit into your own organization's hierarchy.

Understanding how technology affects organizations

Introducing technology into any organization changes the way people work within the organization. The introduction of the first personal computer into the office environment changed the way office assistants prepared documents. The introduction of the cellular telephone allowed people to "take the office on the road." When e-mail was introduced, information moved at the speed of light and allowed documents to be shared instantly. In many cases, these changes began as rather innocuous curiosities. Today, people are connected through local area networks and the Internet, and employees are inextricably linked to job-related software to do their jobs. In short, people work differently than they did before the computer was incorporated into their daily work.

Introducing GIS changes an organization in the following ways:

- ✔ **Change in priorities:** While the GIS becomes more functional, especially when it improves productivity, the priorities of the organization may shift to take advantage of these improvements.

- ✔ **Change in the organizational hierarchy:** While the system grows, the GIS may become more important to organizational groups you hadn't anticipated. Keeping your organizational diagram current helps you track the various groups and their needs.

- ✔ **Change in workflow:** GIS changes who does what, how they do it, and when it gets done. Tasks that took weeks if performed by hand with analog maps, might take hours with GIS. However, the people doing the work might need completely different skills, which could cause managers to hire different personnel to accomplish the same tasks.

- ✔ **Change in the types and amounts of products:** Because GIS offers different output, such as animations, flythroughs, and many more, the products you generate may augment or even replace your existing ones.

- ✔ **Changes in training needs:** The speed of change is increasing. Both the organization and the individuals may need training to keep up with new developments. You definitely need to have a well-designed plan for training.

- ✔ **Change in financial distribution:** You can't get GIS on the cheap. You may need to spend money on GIS analysts, rather than field personnel.

- ✔ **Change in space allocation and design:** You might want to get rid of some of those map cases and get a good climate-controlled room or two.

Managing people problems

Organizations require people — people to run them, people to supply them, people to finance them, and people to manage them. Well, not surprisingly, GIS also requires people. Good, dedicated, hard-working, well-equipped, happy people can make any operation (including GIS) function at peak efficiency. On the other hand, ineffective, apathetic, lazy, poorly trained, disgruntled people can just as easily destroy any chance for an operation's survival.

Nearly every recent GIS system-wide failure has resulted from people and planning problems, not from poor data, computer issues, software, or earthquakes. Here are a few scenarios that can result in failure:

- ✔ **Ineffective sponsorship:** Some systems fail because the system sponsor loses interest when the organization or governing body doesn't see the value and removes support. This situation might occur because of lack of information, misinformation, or a lack of useful output. Organizations need someone to champion the GIS so that the system sponsor is always aware of its potential and actual successes.

- ✔ **Missing or faulty goals:** A GIS that has no goals (or has goals that are at odds with the overall organization's goals) is almost certain to fail or operate poorly. I've seen more than one system fail because it was built around sources of data (such as new sensors and GPS), rather than with an idea of products or potential clientele.

- ✔ **Incomplete implementation:** Systems that meet only part of the organization's spatial data goals are a failure, even if they keep operating. For example, you may have five units that need spatial products, but only three of them use the GIS. This problem can stem from unit managers not understanding the usefulness of GIS data. Or perhaps certain units or individuals resist the technology because they don't see the usefulness or fear that introducing GIS might compromise their position in the organization.

  I've seen system implementations that only partially meet each goal of the organization. You might have five units that all need GIS output, but none of them gets all the products they require. In many cases, this situation occurs because the system sponsor hasn't allocated the necessary resources to complete the implementation, which results in a lack of trained personnel (or training for existing personnel), insufficient data, or a lack of education about the potential for the GIS.

- ✔ **Lack of cooperation:** Turf wars, particularly within organizations that have multiple GIS-related operations, have caused many GIS implementations to operate poorly. Large, complex, multi-user organizations need strong management when data sharing, equipment access, and jealousy about disparate resource allocation cause internal units to resist cooperating.

✔ **Resistance to change:** Employees, particularly long-time employees who might feel marginalized by the new technology, may drag their feet when it comes to adopting unfamiliar technology. Here's an old and still quite true saying: If they're not part of the solution, they're part of the problem.

People problems aren't unique to GIS operations. One of the biggest issues with incorporating GIS into organizations is that it introduces a massive increase in technological dependence. This new technology causes changes in how work is done and in the expectations of increased productivity. And both technology and change are accelerating.

Planning for integration

Ken Eason devised a series of principles (known, not surprisingly, as *Eason's principles*) that can help you introduce any technology into existing organizational structures:

✔ **Serve the organization's needs, rather than just provide technical support.** Make the integration of GIS part of the organization; don't relegate it to some secondary status, like you would fixing broken computers or reformatting hard drives.

✔ **Give employees the ability and willingness to make the system work.** Employees who can't contribute to the system because they lack the necessary knowledge or ability often become unwilling to support that system.

✔ **Prepare a planned process of change based on how the organization workflow will change.** Because organizations typically have their own structure, any change in that structure should produce as little disruption as possible.

✔ **Make employees stakeholders in the system.** For example, you can tie job responsibility or individual employee objectives to GIS system success. Employees who benefit from innovation will support it.

✔ **Make sure that the system meets the organization's goals or solves a problem.** Incorporating new technology because it's hot, rather than to address existing goals, often causes the organization to stray from its original purpose.

✔ **Provide a system that allows cooperation and make incumbent employees feel like they're a part of that system.** Established, loyal employees are an asset to any organization, so give them every opportunity to become integrated into the new approaches.

✔ **Meet the needs of the individual employees.** If the system makes the workload easier or reduces employees' stress, allowing them to enjoy their work more, that system has a much better chance of success.

✔ **Provide education and training for the organization's management and the individual employees.** Although managers may not be responsible for operating the GIS, they need to understand its relevance to the organization's goals. GIS is advanced technology, and it's always changing. The more employees know about the technology, the more quickly they can understand and fully integrate it.

✔ **Plan a progressive form of evolutionary growth.** Changes in organizational needs, technological innovations, software updates, and the natural growth of the GIS demand that you have planning for such changes in place to avoid disruption in the workflow.

✔ **Complement existing design principles and organizational change methodologies.** The more closely the new approach functions like the old, the more quickly both management and employees will adopt it, leading to a smooth transition.

Looking Before You Leap (And Afterwards, Too)

You can't design and implement a GIS in your organization without a little upfront homework. And you can't maintain successful GIS operations without checking up on how the system is working. But you can accomplish both — the upfront homework and ongoing checkups — by performing some key analyses.

Performing needs analysis

Organizations that plan to adopt GIS don't always know their needs. I've been involved in some design projects in which the organization knew a great deal about what GIS is and how they wanted to use it. More commonly, though, organizations that are considering GIS seldom know their GIS needs, and they also don't understand the potential of GIS to fundamentally change how the organization's operations get done. Most organizations that begin to incorporate GIS benefit from an initial needs analysis. This kind of analysis can get complicated and detailed. Each GIS design consultant seems to have a different approach.

Your GIS implementation has four major areas where you must assess what it can provide for your organization:

✔ **Application design:** Requires that you know exactly what maps, graphics, and other decision-support products you want the software to produce for you. The system designer typically conducts interviews to determine what the specific information products are that might result from the implementation of GIS. Good designers often reverse-engineer your organization to define your application needs. So, they have to identify your current goals and products, and determine how GIS might be able to enhance your operations, instill efficiency, and develop better products.

Make getting your existing products set up for GIS automation priority number one. When your organization figures out more about the capabilities of GIS, you begin asking the system to do more. When your needs increase, ask your vendors to help you respond to them.

✔ **Analytical model design:** Know what analytical techniques you need your GIS to perform. These techniques differ for ecological, economic, military, and any other type of analysis. System designers use their knowledge of the GIS capabilities to guide you through the maze of possibilities.

✔ **Database design:** Various software packages use different data models. The application design needs and the analytical model needs you define for your GIS work together to determine your database needs. A thorough database design involves everything from formatting and gathering data to devising a complete plan for creating and maintaining your GIS database.

✔ **Institutional/system design:** When you implement GIS, you need to consider the people in an organization. You need to answer questions regarding personnel, equipment, space, organizational structure, enterprise-wide goals and objectives, and costs and benefits. Costs and benefits, in particular, often determine whether a GIS is even viable.

Most organizations conduct the process of needs assessment with the assistance of outside GIS design consultants. If you already know which software you want to use, these companies can provide design consultants. Many organizations just starting out with GIS prefer to have a system developed from the ground up so that the design document they receive allows them to create an RFP (request for proposals), which gets GIS companies to beg for your business. Using a design document in an RFP often gets you the system that's right for you, rather than the one that some GIS firm wants to sell you. (See Chapter 23 to find out what questions to ask GIS vendors.)

For some businesses, you can look at your competition to see what systems they use. They may not be forthcoming about the specifics, but the companies that serve them will be quick to let you know that they're around.

Performing a cost/benefit analysis

You need to assess the costs and benefits of a GIS implementation if the GIS is designed to bring real financial benefits to the organization. Although organizations that don't generate income from the implementation of GIS don't use it as often, they might still be able to justify the expense because of other benefits, such as savings in time and improved product quality.

When you perform a cost-benefit analysis (CBA), you determine whether something has more benefit than expense, or vice versa. For example, if you invest $10,000 in something and it returns $16,000, you've netted $6,000 on your investment. Usually, you write this relationship as a ratio of the benefit divided by the cost. In this example, your ratio would be 1.6 to 1.

A common business practice is to look at the benefit-to-cost ratio for any large investment, including software tools like GIS. In some businesses I've worked with, a ratio of 1.6 to 1 would be a minimum standard needed to justify the expenditure.

Although a 1.6 to 1 ratio is good, a ratio of 2.2 to 1 (or even 3 to 1) is better. However, consider other institutional, intangible, or even non-economic considerations that can justify a 1.6 to 1 or even lower ratio. Such considerations might include a desire to shift direction, improve worker satisfaction, or improve product quality. All these changes might prove important in the future.

When calculating economic benefits, remember that future benefits are less valuable than immediate benefits — in real dollars — because of inflation. For this reason, most applications of CBA employ an annually discounted rate, usually 7 percent. So, a benefit of $1,000 today will be worth only $930 next year.

Incorporating GIS into any organization often quickly incurs costs, but enjoying the benefits can take some time, which is one of the more difficult barriers to its adoption. In most cases, the first year's costs outweigh the benefits because the initial outlays involve staffing, planning, setup, automation, application development, and training. One pragmatic approach to CBA for GIS implementation is to use a six-year planning horizon. In the first year, your ratio might be 0.5 to 1, which means it cost you more than you made. Over the years, this ratio should improve. By year six, you should see ratios of 2.5 to 1 or better. Typically, a six-year average exceeds 1.6 to 1 if you design your GIS implementation well.

When performing CBA, you may have trouble defining the units of measure for both costs and benefits. You base your costs and benefits on the costs and financial benefits of the GIS products. Costs, often easier to determine than benefits, include hardware, data, software, application development, staffing, training, and maintenance. The benefits come in four groups:

✔ **Direct:** Improvements over the manual methods employed before GIS. Operational efficiencies often result from reducing staff, increasing staff productivity, and even reducing purchases of analog products.

✔ **Agency:** Increases in the company productivity and product quality.

✔ **Government:** The benefits to government agencies other than the one actually implementing the GIS. This benefit also applies to partnerships that result in reduced costs or shared resources. Better, more timely products result in greater customer satisfaction and often increase sales.

✔ **External:** Improvements in the quality and quantity of external factors, perhaps including the impact of the products (for example, you may save a historic site as a result of the output of GIS), improved public perception of your company (the good-neighbor effect), and the regional benefits of increased tourism. These external benefits may be harder to quantify but often have a lasting, long-term positive effect on your company.

Understanding initial versus ongoing analysis

Organizations often consider GIS design when thinking about a new operation or implementation other than what was originally intended. During this consideration period, have systems analysts review your needs; perform all the necessary cost-benefit analyses; and make initial staffing, education, hardware, software, and all the other recommendations to ensure you're headed in the right direction.

But GIS operations already under way, even those that you design well from the start, can benefit from periodic review and remediation. Systems designers are trained to detect the effects of changes that might reduce the potential of your GIS, keeping it from operating at peak efficiency.

Using Change Detection

You can prevent the slow decay in GIS functionality by putting in place a good system of change detection. Change really is the only constant in the universe — and with the rapid advances in technology, change is only going to happen faster. Most business models assume that the organization is dynamic and subject to change. Hardware, software, and operating systems often change most quickly. But organizations also change. The following sections describe the two major types of change that you need to monitor and adjust to.

Technological change

Computer technology has undergone massive changes — from mainframes to minicomputers to PCs. The operating systems that control the machines also affect the GIS software, the supporting software (such as controllers), and third-party software used to support your GIS operations. You need to be aware of the changes so that you can react to them. And you need to balance the cost of acquiring new technology versus the cost of maintaining technology that's obsolete and no longer supported by new versions of the software.

Institutional change

Institutional change can be steady and constant, but factors such as company mergers, new management, new product needs, and new priorities may occur suddenly and without warning. Any of these circumstances warrant periodic systems review.

Here are a few tips to help you detect and manage change in your organization:

✔ **Create an enterprise plan.** The surest way to know the GIS's overall benefits to the organization is to see where it fits in the overall goals and objectives. Although GIS is often a portion of a larger organization, having a system-wide plan allows you to identify where GIS most benefits the organization, regardless of how it changes.

✔ **Review the product line.** Change almost always results in new products and the abandonment of older, obsolete products. Make sure that you review the priorities of the GIS products periodically and make adjustments, as needed.

✔ **Acquire technology.** Besides ensuring that you have adequate technology at the beginning of any GIS operation, you need to establish a steady flow of acquisition, ensuring that you can jettison obsolete technology and always stay current with industry standards in hardware and software.

✔ **Keep management apprised of changes.** Upper-level management generally doesn't have an "in the trenches" view of changes in personnel, training, computer hardware, or software. To guarantee that your organization meets data, software, training and other needs, management must be constantly aware of these needs. You absolutely must establish methods of communicating to management the organization's changing needs and the associated resources necessary to meet those needs.

✔ **Continue to review your plan.** Besides looking at the enterprise-wide goals and objectives, you really need to reconsider the status of your original plan. Identify which parts of the plan are working smoothly and which need adjustment. You may want to have an outside systems designer take a look at your operation from time to time because GIS design professionals are trained to detect subtle changes you might miss.

Part VI
The Part of Tens

The 5th Wave

By Rich Tennant

"You know most campers just bring a compass with them, not an entire GIS software office suite."

In this part . . .

In some ways, GIS is like a smorgasbord with lots of choices. In this Part of Tens, I offer you a sort of menu that helps you choose among the GIS software packages with different capabilities and price ranges. You'll also want someone to serve you, so I provide you with a list of questions you should ask your vendor. Finally, because you must also have data to feed into your GIS, I show you ten sources of spatial data from all over the world — it's an international smorgasbord!

Chapter 21

Ten GIS Software Vendors

GIS software vendors are as varied as the data you encounter inside the software. These vendors range from companies that specialize in industry-specific software (such as Surfer for terrain analysis and CrimeStat for crime analysis), to more generic software companies (such as Environmental Systems Research Institute, Intergraph, and GE Smallworld). With the list of GIS software vendors tipping the scales at "a whole lot" and growing, you may find having a list of ten well-known vendors helpful. For an updated, comprehensive list of GIS vendors, see www.geog.ubc.ca/courses/klink/resources/gis_vendors.html.

Each of the vendors listed in this chapter has a loyal client base, so it doesn't hurt to ask folks who use each company for guidance when you're choosing the company that's right for you. Most GIS users are more than happy to talk about their experiences.

Environmental Systems Research Institute

Environmental Systems Research Institute (ESRI, at www.esri.com) boasts the largest market share for GIS and has come by that boast honestly. Although their products have a steep learning curve, the power this software provides is worth the time investment:

✔ **Product line:** The primary ESRI product is ArcGIS (shown in Figure 21-1). This product comes in many forms and price lines, including online, server, and desktop versions. It incorporates both vector (polygonal) and raster (grid) data models, and it has a new data model called the geodatabase that's object oriented and allows database designers to incorporate detailed geographic properties into the geographic features. Other products include Business Analyst, Business Map, Route Map, and many more packages for traditional or Internet delivery and software development. You can download a demonstration version of the ArcGIS software at `support.esri.com/downloads/`.

One of the ArcGIS extensions, ArcGIS 3D Analyst, has added a new technology for visualizing extremely large three-dimensional geographic data, from local to global perspective. Called ArcGlobe, this product allows the user to interact seamlessly with any geographic data as map layers on a three-dimensional globe (Figure 21-2). Think Google with underlying GIS data to analyze.

Basic tools of the ESRI software include

- Editing tools

- Cartography tools

- Internet mapping and analysis

- On-the-fly map projections

- Large data-format compatibility

- A full suite of spatial analysis tools

- Spatial statistics and visualization

- Coordinate geometry tools

✔ **Training opportunities:** ESRI provides many forms of training, including instructor-led instruction at its facilities or yours, online courses through its virtual campus, and training modules at its tradeshows and annual user meetings. Through its own publishing house, ESRI provides software documentation, as well as tutorial and training texts. Some of these documents come with educational versions of the software that work for a time and then stop running.

The ESRI Web site includes both educational and research-focused audio podcasts that you can download. For educators, ESRI provides a student version of the software that allows students to become familiar with GIS concepts by using ESRI's professional products.

✔ **Other services:** Far more than a GIS software vendor, ESRI provides many consulting opportunities, such as custom implementation and assistance in the design and application of its software. ESRI also provides datasets and support services for your installation and Web hosting for online delivery of GIS analysis and inquiry.

✔ **Platforms:** The ESRI GIS software works on Unix- and Windows-based machines. The online products require Internet Explorer, Firefox, or other Web browser software to run.

✔ **Contact information:** Corporate Headquarters: 380 New York St., Redlands, CA 92373-8100. Phone: 909-793-2853. Web: www.esri.com.

Figure 21-1: Presenting ArcGIS.

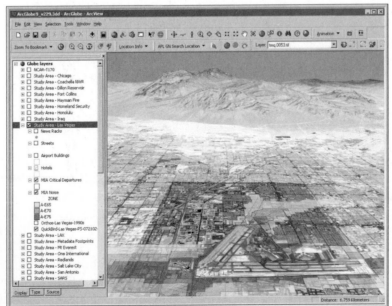

Figure 21-2: What you see on ArcGlobe.

PitneyBowes MapInfo Incorporated

This well-established company uses "locational intelligence" as its motto — it focuses on the use of maps to improve decision-making, even for non-GIS users. Its primary selling point has often been its ease of use:

- ✔ **Product line:** The primary product is MapInfo Professional (Figure 21-3), a general-purpose software package. Primarily a vector-based package, it possesses raster capabilities, especially raster display. Other products in the PitneyBowes line include Internet server packages, software embedding packages for your existing applications, and application development software.

  MapInfo capabilities include

    - Detailed mapping

    - Identification of patterns and trends

    - Extensive data analysis

    - Demographic analysis

    - Remote database support

- ✔ **Training opportunities:** PitneyBowes offers both public and private training. Private training includes five types of classes:

    - Standard courses

    - Getting-started courses

    - Á la carte courses

    - Practical application days

    - Custom development courses

- ✔ **Other services:** You can use the PitneyBowes applied research services for business (the primary use), consulting (including installation, implementation, design, and custom application development), and Web hosting.

- ✔ **Platforms:** MapInfo requires a Microsoft Windows (XP or Vista) operating system.

- ✔ **Contact information:** Corporate Headquarters: One Global View, Troy, NY 12180. Office: 518-285-6000 or 800-327-8627. Fax: 518-285-6070. E-mail: sales@mapinfo.com. Web: www.mapinfo.com.

Figure 21-3:
MapInfo
Professional.

Intergraph

Once known primarily for its powerful Automated Mapping and Facilities Mapping software, Intergraph provides geospatial solutions primarily in security, government, and infrastructure industries:

✔ **Product line:** Intergraph's premier GIS software package is called GeoMedia Professional, a vector-based GIS product (see Figure 21-4). Many special-purpose modules include GeoMedia Terrain for working with surfaces, Grid for raster modeling, Image for working with scanned imagery or satellite data, and WebMap Professional for presenting maps online, in addition to data server and other products.

Major capabilities of the software include

- Queries
- Buffer zones
- Thematic mapping
- Spatial and arithmetic functions

- What-if scenario building

- Web-based map visualization

- On-the-fly map projections

✔ **Training opportunities:** Intergraph offers hands-on training at its facilities, at your facility, and at its annual conference. The company has a Registered Research Laboratory program that universities can use, and Intergraph increasingly encourages academic use of its software by providing it free for use in classes. A 52-week license of GeoMedia Professional is available for students to use at home.

✔ **Other services:** Intergraph has a long tradition of fitting its product line to your particular solution. Even the Web site (at www.intergraph. com) is solution-driven, rather than product-driven. You can find consulting, design, and implementation services on Intergraph's Web site, as well as technical support for its product line.

✔ **Platform:** Intergraph's software works on a Microsoft Windows operating system (XP or Vista).

✔ **Contact information:** Intergraph Corporation, P.O. Box 240000, Huntsville, AL 35813; or 170 Graphics Dr., Madison, AL 35758. Phone: 256-730-2000 or 800-345-4856. Fax: 256-730-2048. Web: www.intergraph. com (you can find an e-mail contact form on the Web site).

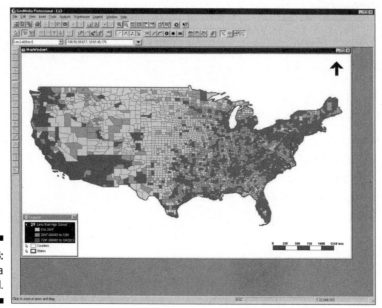

Figure 21-4:
GeoMedia
Professional.

Clark Laboratories

The laboratories of Clark University have a powerful but inexpensive package called IDRISI that performs both GIS and remote sensing image processing tasks. Its current incarnation is called IDRISI Andes.

- **Product line:** IDRISI is the core product available from Clark Laboratories, but a recent parallel product called Land Change Modeler provides additional functionality and works in tandem with the ESRI ArcGIS product. Some of these Clark Laboratories products' prominent features include

  - Land-change modeling

  - Time series analysis

  - Image processing

  - Image classification

  - Neural networks

  - Decision support

  - 2D and 3D visualization

  - Surface analysis

  - Spatial statistics

- **Training opportunities:** Comprehensive tutorials are included with the software. University faculty often use IDRISI tutorials as a teaching tool for both GIS and remote sensing courses because of ease of use and price. Clark Laboratories offers academic pricing at 50 percent of retail cost, as well as $95 student licenses.

- **Other services:** Clark Laboratories provides technical assistance for software installation, operation, data compatibility, and selection of solution-specific modules. Although it interacts with agencies to perform analyses with its products, it is not focused on consulting.

- **Platforms:** IDRISI employs the Microsoft Windows platform.

- **Contact information:** Clark Labs, Clark University, 950 Main St., Worcester, MA 01610-1477. Phone: 508-793-7526. Fax: 508-793-8842. E-mail: idrisi@clark.edu. Web: www.clarklabs.org.

Autodesk, Inc.

Best known for its computer-aided drafting (CAD) software, AutoCAD, Autodesk now sells a line of geospatial products that are designed to act as a bridge between CAD and GIS:

- **Product line:** Autodesk's primary geospatial product line is called, not surprisingly, Autodesk Geospatial. It consists of a number of component products. AutoCAD Map3D is designed to streamline the geospatial data input and management tasks. Other component products assist in online data sharing and team collaboration, incorporation of scanned imagery, data conversion, coordinate geometry, and other management tasks.

- **Training opportunities:** Autodesk has several authorized training centers that deliver instructor-led software training. You can also find a line of training manuals available from the company. These manuals come in three types: Autodesk official training courseware (AOTC) used in the training centers, Autodesk official certification courseware for those who want to become certified to teach Autodesk's software, and Autodesk authorized training courseware produced by Autodesk partners. The courses come in four levels: essentials, transition, advanced, and solutions.

- **Other services:** Autodesk provides technical support and offers consulting services. Consulting mostly focuses on the use of the existing products, rather than one-of-a-kind application development.

- **Platforms:** AutoCAD Map 3D and associated products require the Microsoft Windows operating system. Autodesk offers Web-based tools for UNIX, Windows, and Mac operating systems.

- **Contact information:** Autodesk, Inc., 111 McInnis Pkwy., San Rafael, CA 94903. Phone: 415-507-5000. Fax: 415-507-5100. Web: http://usa.autodesk.com (you can find an e-mail contact form on the Web site).

GE Smallworld

Smallworld, once its own company, is now part of General Electric Energy Corporation and provides a suite of geospatial tools to support communications, utility, and public systems organizations:

- **Product line:** The core product for GE Smallworld is called the Core Spatial Technology product suite. It enables encapsulated, reusable components for rapid development and reduced custom code generation. Strong features of the software are its scalable architecture and seamless open data access.

Smallworld's software capabilities include

- Data access and integration

- Thematic mapping and visualization

- Sketching and measurement

- Querying

- A complete set of spatial analysis tools (network, buffering, polygon, cluster, and proximity analysis)

✔ **Training opportunities:** GE Energy offers courses for engineers, developers, and administrators who use GE's suite of products at its corporate sites. You can schedule additional locations, times, and courses if sufficient demand exists; you can even hold these classes at your own organization's location.

✔ **Other services:** As part of a larger corporation, Smallworld provides consulting, installation, and customization of its product line, as well as product technical support.

✔ **Platform:** UNIX and Microsoft Windows.

✔ **Contact information:** GE Energy, 2000 S. Colorado Blvd., Ste. 2-1100, Denver, CO 80222. Phone: 303-779-6980. Fax 303-779-1051. Web: `www.gepower.com` (use the e-mail contact form on the Web site to send a message).

PCI Geomatics

A full-service geospatial company, PCI Geomatics incorporates remote sensing, GIS, *photogrammetry* (the science of exact measurement of features on aerial photography), and cartography. Originally focused on analyzing remotely sensed images, its new product line offers many opportunities for geospatial analysis and display:

✔ **Product line:** PCI's core product, Geomatica, now at version 10, is a desktop product that has both raster and vector capabilities, and boasts support for more than 100 spatial data formats:

- Geometric correction of imagery

- Projection and re-projection

- Data visualization

- Image classification

- Robust geospatial analytics

- Photogrammetric functions

- Cartographic production

- **Training opportunities:** Besides an included set of tutorials, PCI offers on-site or customized training sessions. These sessions can be tailored to meet specific needs.

- **Other services:** PCI provides on-call technical support by phone or e-mail, software updates, data resources (including a line of data-handling tools), and a discussion forum. It also has an extensive list of white papers and articles that describe various applications of its software.

- **Platform:** A desktop Microsoft Windows environment.

- **Contact information:** PCI Geomatics, 50 W. Wilmot St., Richmond Hill, Ontario, Canada L4B 1M5. Phone: 905-764-0614. Fax: 905-764-9604. E-mail: info@pcigeomatics.com. Web: www.pcigeomatics.com.

Leica Geosystems

The motto at Leica is "Powering Geospatial Imaging." This motto reflects its original focus on digital image processing software. Over the years, Leica has grown to incorporate a robust raster-based GIS suite, as well as a suite of photogrammetry tools:

- **Product line:** The primary product, ERDAS IMAGINE, provides a wide array of tools for geospatial analysis:

  - Import/export

  - Data preparation

  - Map composition

  - Image enhancement

  - Image classification, including sub-pixel classification

  - Raster GIS modeling

  - Stereo analysis

  - Batch processing

- **Training opportunities:** Leica offers a full suite of training opportunities, including corporate classroom training at its corporate headquarters, on-site training, and Webinars. Leica also offers a free demo version of its software for potential clients.

- **Other services:** Primarily a software vendor, Leica provides technical support via e-mail and the Internet.

- **Platform:** A desktop Microsoft Windows environment.

- **Contact information:** Leica Geosystems Geospatial Imaging, LL, 5051 Peachtree Corners Circle, Norcross, GA 30092-2500. Phone: 770-776-3400 or 866-534-2286. Fax: 770-776-3500 (general inquiries). E-mail: info@gl.leica-geosystems.com. Web: www.leica-geosystems.com.

Bentley GIS

Bentley Systems is a leading provider of solutions to assist engineers, architects, contractors, governments, institutions, utilities, and private owners with designing, building, and operating infrastructures (including buildings, bridges, and utilities). One major tool for mapping, planning, and designing these solutions is its GIS software.

- ✔ **Product line:** Microstation GeoOutlook is a spatial decision support system for mapping and GIS analysis. It has tools that can provide data visualization, map measurement, dimensioning, database access, and plotting. Other major features include

  - Data collection

  - Data editing

  - Cleanup and analysis tools

  - Integrated engineering and mapping on a single platform

- ✔ **Training opportunities:** Bentley offers online educational materials and classroom training at its locations worldwide. Bentley also offers account-specific training at your location.

- ✔ **Other services:** The Bentley Institute provides consulting services. The company also offers services for integration and software implementation. The Managed Solutions branch provides software-plus-services packages for clients.

- ✔ **Platform:** A desktop Microsoft Windows environment.

- ✔ **Contact information:** Bentley Corporation, 685 Stockton Dr., Exton, PA 19341. Phone: 800-236-8539. Outside the U.S.: 610-458-5000. Web: www. bentley.com.

GRASS GIS

Originally developed by U.S. Army Construction Engineering Research Laboratories, GRASS (Geographic Resource Analysis Support System) remains the most powerful and possibly the most common open source GIS package on the market. Reflecting its natural-resources–analysis roots, this package contains both GIS and remote sensing image processing capabilities:

- ✔ **Product line:** GRASS is a single product, with many modules. It is has both raster and vector data models. Its major capabilities and features include

  - 2D raster analysis and 3D volume management

  - 2D and 3D vector models with SQL DBMS capabilities

- Image processing

- Network analysis and referencing

- Visualization in both two and three dimensions

- Batch processing

- Standard raster and vector data format interoperability

✔ **Training opportunities:** Because GRASS is open source software, no one offers formal training — but you can access a large user community that can readily provide help.

✔ **Other services:** The Open Source Geospatial Foundation's own mission statement says it all — its purpose is to "support the development of open source geospatial software, and promote its widespread use." It doesn't provide consulting services or technical support.

✔ **Platform:** GRASS is a multi-platform software solution. It operates on computers running Unix, Microsoft Windows, or Mac OS X.

✔ **Contact information:** Open Source Geospatial Foundation, 14525 S.W. Millikan, #42523, Beaverton, OR 97005-2343. Phone: 250-277-1621. E-mail: tmitchell@osgeo.org. Web: www.osgeo.org.

Chapter 22

Ten Questions to Ask Potential Vendors

In This Chapter

▶ Identifying the right vendor for your needs

▶ Understanding the products and services that vendors provide

*F*inding a vendor that meets — and even exceeds — your needs is critical to ensuring your success when working with GIS. In order to find that vendor, you need to know what questions to ask. This chapter provides a good starting point for your decision-making process. Whatever you do, don't rush your decision.

What Services Do You Offer?

Your first questions for a vendor should be based on your needs. Figure out what you require for output (called Spatial Information Products). Your output requirements give you an idea of the software capabilities, data structures (for example, raster or vector), and analytical capabilities that you need. Also, determine what, beyond the software, you'll need from the company. Do you need consulting, implementation and setup, training, complete solutions, individualized services, or data?

Ask yourself what you currently use the GIS for in your organization so you can be sure the software meets those immediate needs.

Can You Show How Your Product Will Meet My Needs?

Among the most powerful methods of evaluating the utility, quality, and reliability of a vendor is to view examples of products and services that it provided to other clients. Reputable companies provide examples from their own portfolios, as well as the names and contact information of past clients.

You might want to ask the potential vendor to provide several names of satisfied and dissatisfied clients. A large amount of resistance to supplying this information tells you something about the vendor's lack of confidence in its products and services.

What Data Formats Does Your Product Support?

Of the many different spatial data formats, some are standard and readily available, but others are proprietary. If you know your data format needs, include them in any request for proposals. Of course, you probably know some — but not all — necessary data formats, so ask the vendor for a complete list of data formats that it can either provide or can convert to and from. You don't necessarily want to go with the largest list because that list may not include important formats. Ask the vendor to provide not just a list of formats, but some descriptive information regarding what those formats are and who created them.

How Do You Handle Communications and Change Requests?

Projects aren't static. While your needs and expectations change, you want a vendor who can adapt to your changing needs. So, make sure that you can have constant communication with the vendor. Ask who will communicate with you, how he or she will communicate, and how often.

The more dynamic your operation, the more you need the assurance of communication. Establish this communication protocol right from the start to avoid any misunderstandings.

What Hardware Expertise Do You Have?

Your computer and its operating system are the lifeblood of your GIS. Additionally, you need peripherals such as monitors, input devices, plotters, printers, and even components such as storage and memory. A vendor that has expertise in finding solutions for a wide variety of hardware configurations provides yet another level of security and reliability for your GIS operations.

Ask the potential vendor for specific services and expertise regarding different platforms, operating systems, and peripherals under differing implementation circumstances. If, for example, you need to use your GIS in hazardous environments, you might ask about what *hardened* equipment (equipment that can withstand rough treatment encountered in field or even battlefield conditions) you might need.

What Does the Price Include?

You absolutely need to know exactly what you get from the vendor for the price you pay. If, for example, the vendor fails to include all the GIS parts and services when bidding the job, who's going to pay for any cost overruns that may occur? Does the price include hardware, software, installation, database development, upgrades, technical support, or other services?

How Long Until the System Is Operational?

If your tasks are time-critical, be very specific about when you need final delivery of the product. More importantly, perhaps, you want to know when the system will be fully implemented and operational. Just because the hardware and software are in place doesn't mean that you're ready to use them. You may need to go through training and data conversion, or even module creation and testing, which require additional time. Ask for a schedule of progress reports, as well as some formal agreed-upon procedures if the vendor doesn't complete the project on time.

What Happens If the System Crashes?

Every system occasionally experiences software crashes or hardware failures. Most GIS vendors have some form of maintenance contract. Be sure to evaluate the vendor's specific responsibilities if you experience software or hardware issues. Can the vendor provide on-site assistance, or phone and/or Internet support? Whether the vendor provides hardware support, as well as software support, may dictate the type of maintenance contracts it offers.

What Are Your Quality-Control Procedures?

Quality control is vital at all stages of the GIS life cycle. Find out whether the vendor has an existing set of quality-control procedures that ensures you get the product you need in the time required.

Good companies have written quality-control procedures. Ask to see a copy of these procedures. You may also want to ask who's in charge of those procedures and get that person or department's complete contact information.

What Are Your Performance Guarantees?

While your GIS activities progress, you're almost surely going to encounter shortcomings and mishaps along the way. Because you'll be signing a written contract with the vendor you choose, ask each vendor about its policy regarding restitution or solutions if the GIS or consulting services don't meet your expectations. Good GIS vendors want to build a continuing relationship with their clients so they can continue to do business with them. To keep you in their client base, they usually do their best to make you happy.

Find out what a vendor would do under certain circumstances, for example, how it handles a request for changes in contracts as your organization grows or a request for rapid response for technical help during critical times in a project. If the vendor is reputable, it may have real-life stories that it can relate to your questions, demonstrating both its professionalism and its good faith.

Chapter 23

Ten GIS Data Sources

*T*he driving force behind all GIS analysis is data. GIS data come in many forms, from many sources, for many uses, with many scales and levels of accuracy, in different datums and projections, and at a huge variety of prices. Depending on your application and your specific needs, you may want to develop your own datasets. Developing your own datasets allows you to have total control over the content and accuracy, not to mention the ownership, of your data. Industry experts have claimed that data conversion (creating digital databases) often accounted for as much as 60 percent of the cost of system development. Large operations faced an enormous cost. Although these percentages have decreased over the years, they are still a substantial portion of the cost of operating a GIS. And so you might want to examine the possibility of obtaining datasets that already exist.

The sections in this chapter give you an overview of some handy GIS data sources, and Table 23-1 outlines their important features. While you're evaluating sources, here are some questions that you can ask yourself to help you determine which tool you want to use:

✔ Does your GIS have metadata management tools?

✔ Do you need a tool that supports content standards (specifically the Content Standard for Digital Geospatial Metadata, or CSDGM)?

✔ Do you need document data resource beyond what your GIS does?

✔ Do you need a distributable tool for your partners?

✔ What features are most important to you? For example, you may consider auto-capture of information, bundling metadata with data, creation and use of templates, user interface, or robust help menus and tutorials important features.

Table 23-1		Features of GIS Data Sources		
Data Source	**Type / Format**	**Data origin / Compliance**	**Access / Price**	**Coverage**
GIS Data Depot	General Various themes and scales / Industry-standard formats	Original datasets (for example, U.S. Geological Survey) and value-added translations / SDTS* compliant	Download from user account / Free or priced per data block / Data on CD-ROM at a fee	Worldwide
Environ-mental Systems Research Institute (ESRI)	General Various themes and scales / Shapefiles, export files, ArcGIS compatible, and image data	Government and commer-cial providers compiled in-house / FGDC**/SDTS compliant; metadata conform to ISO*** 19115	Direct download or on DVD / free, free with software purchase, or separate purchase	Worldwide
National Geospatial Data Clearing-house	General Various themes and scales / Typical image, export, and shapefiles	Government and some third-party / SDTS compliant	Online search and download / free or at cost	United States
Center for International Earth Science Information Network (CIESIN)	General to specific Various themes and scales / ESRI shape-files, grid data, and other industry-standard formats	Government, commercial, and academic providers, as well as value-added products / SDTS compliant	Online search and download / free or restricted for specific datasets; policy doc-uments and metadata catalog search also online	Worldwide

Data Source	Type / Format	Data origin / Compliance	Access / Price	Coverage
Go-Geo!	General to specific Various themes and scales / Formats vary by data provider	International portals, government archives and consortia, commercial, and academic / SDTS compliance dependent on data provider	Redirects to other sites for data access / Method and cost depends on data provider	U.K. with spotty worldwide data
Instituto National de Estadistica Geographia e Informatica (INEGI)	Mostly general, various natural themes and scales / Vector and raster, industry compatible	Mexican government / Agencies compliant	Online search and telephone consultation / Free or per-product pricing	Mexico
CGIAR++ Consortium for Spatial Information	General to specific, various themes and scales / Various, often unique formats from servers and data warehouses	Government, private industry, scientific research, and conservation groups / SDTS compliance unknown	Variable based on member organizations	Worldwide
Australian Consortium for the Asian Spatial Information and Analysis Network (ACASIAN)	Small-scale administrative and infrastructure databases / One or more ArcGIS, ArcInfo, ArcView, and MapInfo formats	Internal ACASIAN products / SDTS compliance and metadata specific to datasets	E-mail an FTP online or CD, Zip disks, tape, other media / Licensing fee varies by user type	Asia and the former Soviet Union

(continued)

Table 23-1 *(continued)*

Data Source	Type / Format	Data origin / Compliance	Access / Price	Coverage
Geoscience Australia	Geodetic, topographic, thematic data-sets / ArcInfo, ArcView and MapInfo	Government-generated datasets / Compliant with Australian standards organization	Online search and download / Free with registration or at cost for optical formats	Australia
Canada Geospatial Data Infra-structure	General to specific Atlas and land-use Various themes and scales / Raster and vector datasets with separate metadata files (XML and HTML)	U.S. and Canadian government agencies / SDTS compliant	Online search and download / Free	Canada

*SDTS: Spatial Data Transfer Standard
**FGDC: Federal Geographic Data Committee
***ISO: International Standards Organization
++CGIAR: Consultative Group on International Agricultural Research

Whichever tool you choose, remember that metadata are essential for maintaining your own databases and obtaining compatible data from external sources.

GIS Data Depot

GIS Data Depot (http://data.geocomm.com) provides free spatial data, as well as spatial data that you can purchase. The Web site makes it clear right from the start that you need to be proficient in GIS and have access to GIS software. You can download datasets, or for a charge, the data provider can produce datasets as CD-ROM media.

In addition to data, GIS Data Depot also provides an array of utilities, translators, viewers, and scripts to assist you in your purchase and subsequent use of the data.

Contact Info: MindSites Group, LLC, 1161 John Sims Pkwy. E., Niceville, FL 32578. Phone: 850-897-1002. Fax: 850-897-1001. Web: `http://data.geocomm.com` (you can find an e-mail form on the Web site).

Environmental Systems Research Institute

Environmental Systems Research Institute (ESRI), besides being a huge player in the GIS software and consulting industry, also provides a wide array of spatial data for use with its many products. ESRI provides the data on DVDs that include HTML help systems that have information about redistribution and a complete set of metadata. ESRI also has an ArcGIS Online Content Sharing Program to allow organizations to share spatial data.

ESRI provides its own ArcGIS data appliance with pre-rendered U.S. and worldwide data residing on its own ArcGIS Server. These data appliances are already optimized for publishing with ArcGIS Server and include imagery, street maps, shaded relief, and elevation data. You can add your own data to this pre-formatted server to reduce computation time and optimize workflow.

Contact Info: Environmental Systems Research Institute, 380 New York St., Redlands, CA 92373-8100. Phone: 909-793-2853. Fax: 909-793-7070. Web: `www.esri.com` (access the e-mail form on the Web site).

National Geospatial Data Clearinghouse

A direct outcome of the United States' Federal Geographic Data Committee's (FGDC) collaborative activity was the establishment of a group of about 250 cooperating government bodies in the United States called the National Geospatial Data Clearinghouse (NGDC). Organizations must apply for membership in the network and satisfy the FGDC standards to belong. The EROS Data Center (a USGS-run geospatial data facility) and the FGDC host the clearinghouse data.

The NGDC provides well-organized and complete metadata. You can easily access that metadata — but NGDC doesn't commonly offer additional tools.

You can choose from several user interfaces to search for data to download. One really nice feature of this interface is the search wizard that allows you to search for geospatial data by category, as shown in Figure 23-1. After you select the category of data that you're looking for, you can narrow the search by selecting a state from a drop-down menu, graphically choosing an area on the search window, or defining the latitude and longitude extent (see Figure 23-2).

Contact Info: MindSites Group, LLC, 1161 John Sims Pkwy. E., Niceville, FL 32578. Phone: 850-897-1002. Fax: 850-897-1001. Web: www.geocomm.com (send e-mail by using the site's e-mail form).

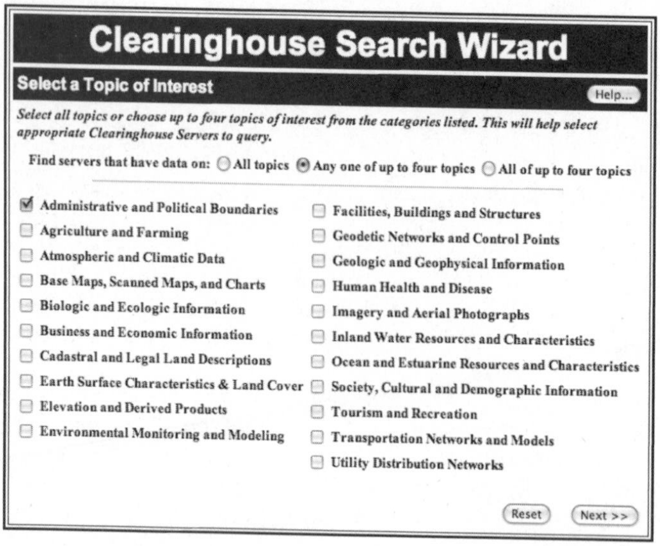

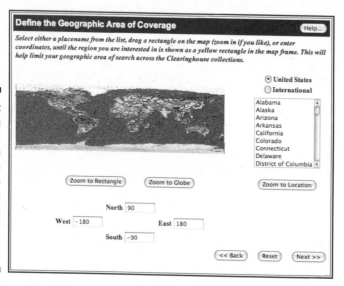

Figure 23-1:
The NGDC search wizard.

Figure 23-2:
Defining the specific locations that you want to search by using the NGDC search wizard.

Center for International Earth Science Information Network (CIESIN)

CIESIN, housed and operated by Columbia University's Earth Institute, provides data, education, support, and research on data integration. Its interests focus primarily on human and environment interactions, which give CIESIN a decidedly more focused perspective on the datasets it provides and on its mission. The primary goals include examining the human dimensions of global change.

Beyond performing its own research and providing valuable data, CIESIN also specializes in digital archiving, metadata systems, and customer relations management, including usability outreach. It has metadata specialists available to ensure that the metadata are complete and correct.

Contact Info: Center for International Earth Science Information Network (CIESIN), 61 Rte. 9W, P.O. Box 1000, Palisades, NY 10964. Phone: 845-365-8988. Fax: 845-365-8922. E-mail: ciesin.info@ciesin.columbia.edu. Web: www.ciesin.org.

Go-Geo!

Go-Geo! is a United Kingdom (U.K.) geospatial data port that focuses predominantly on the academic community, but also serves many others. Funded by the Joint Information Systems Committee (JISC) and operated by the University of Essex and the University of Edinburgh, Go-Geo! provides an opportunity to discover, locate, and retrieve data that would otherwise be difficult to find. Moreover, this initiative also supports the concept of maintaining high spatial-data standards and quality through metadata development.

Go-Geo! provides a number of helpful resources, including terminology, explanations of data types, and descriptions of data standards. In addition, it links to other resources that you can use to conduct data searches. As a soft-funded project (funded by research grants that may not last), Go-Geo! is a trial database provider and may be temporary.

Contact Info: Contact with Go-Geo! is limited to the Internet: www.gogeo.ac.uk. It does have a log-in function for members of U.K. higher education.

Instituto National de Estadistica Geographia e Informatica (INEGI)

INEGI, the National Institute of Statistics, Geography, and Data Processing for Mexico, is a government body that collects and organizes statistical, geographic, and economic information about the country.

INEGI provides consultation services (such as digital map preparation), online consultation, and sale of orthophotos. Agrarian communities have access to additional cartographic services.

Contact Info: Phone: (Spanish language) 01-800-111-46-34 or 449-9-1-53-00, ext. 4797. Web: www.inegi.gob.mx (you can find a Web-based e-mail form on the site).

CGIAR Consortium for Spatial Information (CGIAR-CSI)

CGIAR is an initiative of scientists who have a common interest in international agricultural research. With 15 centers worldwide, the mission is to apply and advance geospatial sciences, primarily for international sustainable agriculture, natural resource management, conservation of biodiversity, and the alleviation of poverty in developing countries.

You can find information about spatial data workshops, metadata resources, intellectual property rights, user directories, and even job and internship offerings.

Contact Info: Robert Zomer, Ph.D., Global Coordinator — CSI-CGIAR CGIAR Consortium for Spatial Information. E-mail: csi@cgiar.org. Also, International Water Management Institute (IWMI), P.O. Box 2075, Colombo, Sri Lanka. Phone: 94-11-2787404. Fax: 94-11-2786854. Web: http://csi.cgiar.org.

Australian Consortium for the Asian Spatial Information and Analysis Network (ACASIAN)

ACASIAN is an applied academic organization interested in the use of GIS databases for Asia (primarily China) and the former Soviet Union. The datasets are licensed and for sale, and ACASIAN encourages collaborative research that uses its datasets. ACASIAN can provide datasets that include special projections for additional fees.

Contact Info: ACASIAN, Lawrence Crissman, Dept. of International Business and Asian Studies, School of Business, Griffith University, Brisbane Q 4111, Australia. Phone: 61-7-3875-7285. Fax: 61-7-3875-5111. E-mail: `l.crissman@ griffith.edu.au`. Web: `www.asian.gu.edu.au`.

Geoscience Australia

Part of the Australian Government's Department of Resources, Energy, and Tourism, Geoscience Australia produces geoscientific information and knowledge. The primary applications are related to decision-making for resource exploration, environmental management, and maintaining the infrastructure critical for the well-being of Australian citizens.

Geoscience Australia provides hardcopy maps and aerial photography, in addition to its digital products. It also has educational materials available for primary and secondary schools. You can also download free data viewers.

Contact Info: Street address: Cnr Jerrabomberra Ave. and Hindmarsh Dr., Symston, ACT 2609, Australia. Postal address: GPO Box 378, Canberra, ACT 2601, Australia. Phone: 1800-800-173 (products and services) or 61-2-6249-9111 (all other inquiries). Fax: 61-2-6249-9999. Web: `http://www.ga.gov.au/`.

Canada Geospatial Data Infrastructure

The Canada Geospatial Data Infrastructure (CGDI) is a combination of technology, geographic data standards, methods, and protocols needed to coordinate and manage Canadian geospatial databases. The databases include topographic maps, aerial photography, aeronautical and nautical charts, satellite imagery, and environmental data (including forestry, soil, marine, and biodiversity inventories). They also contain socioeconomic and demographic (Census) data, and electoral boundaries.

The primary purposes for maintaining this managed resource include geospatial analyses to benefit public health, safety and security, the environment, and Canada's Aboriginal peoples.

Contact Info: GeoGratis Client Services, Natural Resources Canada, 2144 Kind W. Street, Ste. 010, Sherbrooke, Quebec J1J 2E8, Canada. Phone: 613-995-0947. Fax: 819-564-5698. E-mail: `geoinfo@nrcan.gc.ca`. Web: `http://geogratis. cgdi.gc.ca`.

A cautionary note

Not all GIS data are created equal. Commercial firms sometimes package data that you can find for free or at a minimal reproduction fee elsewhere, claiming that the data they offer are value-added data. Calling something value-added implies that the company is justified in charging for these products because it has modified the data in some way beyond their original form.

But you need to determine whether the added value warrants the added cost. The companies may claim that they improved the data's quality, packaged them specifically for your application, or actually added some attributes to the data that make the data considerably different than their raw, unmodified form. Some modifications actually degrade the original data. In some cases, a company really does add necessary attributes, package the data with your specific application in mind, and provide additional services that make the data much more adaptable to your organization. So, you just need to decide whether these changes really add value to the data for you.

Here are some guidelines that I like to use to help me decide whether to buy value-added data:

✔ **Ask for complete metadata.** You can use the metadata to determine the original source of the data and how they were modified, as well as get an indication of the quality of the data. If the company isn't forthcoming with metadata, walk away.

✔ **Ask for a list of clients.** Get contact information for clients who have purchased similar value-added data from the data provider, including both happy and not-so-happy clients.

✔ **Talk to some GIS professionals and ask their advice.** Many universities have GIS faculty who know about alternate data sources or lower-cost data, or they may be able to advise you about the possible value of value-added data. Also, go to user group blogs and lists online to find additional insights about each data provider and the quality of its products.

✔ **Get a contract.** Make sure it includes a guarantee that the data aren't already available free or at minimal cost in essentially the same quality and format that the vendor is offering. The contract should include a statement that explicitly indicates the value that the company added to existing data, especially public-domain data.

✔ **Resist expediency.** You might find the idea of purchasing data that you can obtain quickly very attractive, with the promise of getting your GIS up and running in no time. But remember that the value of your GIS operation lies fundamentally on the quality of your data. If you get poor-quality data quickly, the time it takes to clean up the mess can ultimately cost you more than taking the time to obtain quality data.

Index

• S •

Notes

Notes

BUSINESS, CAREERS & PERSONAL FINANCE

Accounting For Dummies, 4th Edition*
978-0-470-24600-9

Bookkeeping Workbook For Dummies†
978-0-470-16983-4

Commodities For Dummies
978-0-470-04928-0

Doing Business in China For Dummies
978-0-470-04929-7

E-Mail Marketing For Dummies
978-0-470-19087-6

Job Interviews For Dummies, 3rd Edition*†
978-0-470-17748-8

Personal Finance Workbook For Dummies*†
978-0-470-09933-9

Real Estate License Exams For Dummies
978-0-7645-7623-2

Six Sigma For Dummies
978-0-7645-6798-8

Small Business Kit For Dummies, 2nd Edition*†
978-0-7645-5984-6

Telephone Sales For Dummies
978-0-470-16836-3

BUSINESS PRODUCTIVITY & MICROSOFT OFFICE

Access 2007 For Dummies
978-0-470-03649-5

Excel 2007 For Dummies
978-0-470-03737-9

Office 2007 For Dummies
978-0-470-00923-9

Outlook 2007 For Dummies
978-0-470-03830-7

PowerPoint 2007 For Dummies
978-0-470-04059-1

Project 2007 For Dummies
978-0-470-03651-8

QuickBooks 2008 For Dummies
978-0-470-18470-7

Quicken 2008 For Dummies
978-0-470-17473-9

Salesforce.com For Dummies, 2nd Edition
978-0-470-04893-1

Word 2007 For Dummies
978-0-470-03658-7

EDUCATION, HISTORY, REFERENCE & TEST PREPARATION

African American History For Dummies
978-0-7645-5469-8

Algebra For Dummies
978-0-7645-5325-7

Algebra Workbook For Dummies
978-0-7645-8467-1

Art History For Dummies
978-0-470-09910-0

ASVAB For Dummies, 2nd Edition
978-0-470-10671-6

British Military History For Dummies
978-0-470-03213-8

Calculus For Dummies
978-0-7645-2498-1

Canadian History For Dummies, 2nd Edition
978-0-470-83656-9

Geometry Workbook For Dummies
978-0-471-79940-5

The SAT I For Dummies, 6th Edition
978-0-7645-7193-0

Series 7 Exam For Dummies
978-0-470-09932-2

World History For Dummies
978-0-7645-5242-7

FOOD, GARDEN, HOBBIES & HOME

Bridge For Dummies, 2nd Edition
978-0-471-92426-5

Coin Collecting For Dummies, 2nd Edition
978-0-470-22275-1

Cooking Basics For Dummies, 3rd Edition
978-0-7645-7206-7

Drawing For Dummies
978-0-7645-5476-6

Etiquette For Dummies, 2nd Edition
978-0-470-10672-3

Gardening Basics For Dummies*†
978-0-470-03749-2

Knitting Patterns For Dummies
978-0-470-04556-5

Living Gluten-Free For Dummies†
978-0-471-77383-2

Painting Do-It-Yourself For Dummies
978-0-470-17533-0

HEALTH, SELF HELP, PARENTING & PETS

Anger Management For Dummies
978-0-470-03715-7

Anxiety & Depression Workbook For Dummies
978-0-7645-9793-0

Dieting For Dummies, 2nd Edition
978-0-7645-4149-0

Dog Training For Dummies, 2nd Edition
978-0-7645-8418-3

Horseback Riding For Dummies
978-0-470-09719-9

Infertility For Dummies†
978-0-470-11518-3

Meditation For Dummies with CD-ROM, 2nd Edition
978-0-471-77774-8

Post-Traumatic Stress Disorder For Dummies
978-0-470-04922-8

Puppies For Dummies, 2nd Edition
978-0-470-03717-1

Thyroid For Dummies, 2nd Edition†
978-0-471-78755-6

Type 1 Diabetes For Dummies*†
978-0-470-17811-9

* Separate Canadian edition also available
† Separate U.K. edition also available

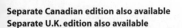
Available wherever books are sold. For more information or to order direct: U.S. customers visit www.dummies.com or call 1-877-762-2974.
U.K. customers visit www.wileyeurope.com or call (0)1243 843291. Canadian customers visit www.wiley.ca or call 1-800-567-4797.

INTERNET & DIGITAL MEDIA

AdWords For Dummies
978-0-470-15252-2

Blogging For Dummies, 2nd Edition
978-0-470-23017-6

Digital Photography All-in-One Desk Reference For Dummies, 3rd Edition
978-0-470-03743-0

Digital Photography For Dummies, 5th Edition
978-0-7645-9802-9

Digital SLR Cameras & Photography For Dummies, 2nd Edition
978-0-470-14927-0

eBay Business All-in-One Desk Reference For Dummies
978-0-7645-8438-1

eBay For Dummies, 5th Edition*
978-0-470-04529-9

eBay Listings That Sell For Dummies
978-0-471-78912-3

Facebook For Dummies
978-0-470-26273-3

The Internet For Dummies, 11th Edition
978-0-470-12174-0

Investing Online For Dummies, 5th Edition
978-0-7645-8456-5

iPod & iTunes For Dummies, 5th Edition
978-0-470-17474-6

MySpace For Dummies
978-0-470-09529-4

Podcasting For Dummies
978-0-471-74898-4

Search Engine Optimization For Dummies, 2nd Edition
978-0-471-97998-2

Second Life For Dummies
978-0-470-18025-9

Starting an eBay Business For Dummies, 3rd Edition†
978-0-470-14924-9

GRAPHICS, DESIGN & WEB DEVELOPMENT

Adobe Creative Suite 3 Design Premium All-in-One Desk Reference For Dummies
978-0-470-11724-8

Adobe Web Suite CS3 All-in-One Desk Reference For Dummies
978-0-470-12099-6

AutoCAD 2008 For Dummies
978-0-470-11650-0

Building a Web Site For Dummies, 3rd Edition
978-0-470-14928-7

Creating Web Pages All-in-One Desk Reference For Dummies, 3rd Edition
978-0-470-09629-1

Creating Web Pages For Dummies, 8th Edition
978-0-470-08030-6

Dreamweaver CS3 For Dummies
978-0-470-11490-2

Flash CS3 For Dummies
978-0-470-12100-9

Google SketchUp For Dummies
978-0-470-13744-4

InDesign CS3 For Dummies
978-0-470-11865-8

Photoshop CS3 All-in-One Desk Reference For Dummies
978-0-470-11195-6

Photoshop CS3 For Dummies
978-0-470-11193-2

Photoshop Elements 5 For Dummies
978-0-470-09810-3

SolidWorks For Dummies
978-0-7645-9555-4

Visio 2007 For Dummies
978-0-470-08983-5

Web Design For Dummies, 2nd Edition
978-0-471-78117-2

Web Sites Do-It-Yourself For Dummies
978-0-470-16903-2

Web Stores Do-It-Yourself For Dummies
978-0-470-17443-2

LANGUAGES, RELIGION & SPIRITUALITY

Arabic For Dummies
978-0-471-77270-5

Chinese For Dummies, Audio Set
978-0-470-12766-7

French For Dummies
978-0-7645-5193-2

German For Dummies
978-0-7645-5195-6

Hebrew For Dummies
978-0-7645-5489-6

Ingles Para Dummies
978-0-7645-5427-8

Italian For Dummies, Audio Set
978-0-470-09586-7

Italian Verbs For Dummies
978-0-471-77389-4

Japanese For Dummies
978-0-7645-5429-2

Latin For Dummies
978-0-7645-5431-5

Portuguese For Dummies
978-0-471-78738-9

Russian For Dummies
978-0-471-78001-4

Spanish Phrases For Dummies
978-0-7645-7204-3

Spanish For Dummies
978-0-7645-5194-9

Spanish For Dummies, Audio Set
978-0-470-09585-0

The Bible For Dummies
978-0-7645-5296-0

Catholicism For Dummies
978-0-7645-5391-2

The Historical Jesus For Dummies
978-0-470-16785-4

Islam For Dummies
978-0-7645-5503-9

Spirituality For Dummies, 2nd Edition
978-0-470-19142-2

NETWORKING AND PROGRAMMING

ASP.NET 3.5 For Dummies
978-0-470-19592-5

C# 2008 For Dummies
978-0-470-19109-5

Hacking For Dummies, 2nd Edition
978-0-470-05235-8

Home Networking For Dummies, 4th Edition
978-0-470-11806-1

Java For Dummies, 4th Edition
978-0-470-08716-9

Microsoft® SQL Server™ 2008 All-in-One Desk Reference For Dummies
978-0-470-17954-3

Networking All-in-One Desk Reference For Dummies, 2nd Edition
978-0-7645-9939-2

Networking For Dummies, 8th Edition
978-0-470-05620-2

SharePoint 2007 For Dummies
978-0-470-09941-4

Wireless Home Networking For Dummies, 2nd Edition
978-0-471-74940-0

OPERATING SYSTEMS & COMPUTER BASICS

iMac For Dummies, 5th Edition
978-0-7645-8458-9

Laptops For Dummies, 2nd Edition
978-0-470-05432-1

Linux For Dummies, 8th Edition
978-0-470-11649-4

MacBook For Dummies
978-0-470-04859-7

Mac OS X Leopard All-in-One Desk Reference For Dummies
978-0-470-05434-5

Mac OS X Leopard For Dummies
978-0-470-05433-8

Macs For Dummies, 9th Edition
978-0-470-04849-8

PCs For Dummies, 11th Edition
978-0-470-13728-4

Windows® Home Server For Dummies
978-0-470-18592-6

Windows Server 2008 For Dummies
978-0-470-18043-3

Windows Vista All-in-One Desk Reference For Dummies
978-0-471-74941-7

Windows Vista For Dummies
978-0-471-75421-3

Windows Vista Security For Dummies
978-0-470-11805-4

SPORTS, FITNESS & MUSIC

Coaching Hockey For Dummies
978-0-470-83685-9

Coaching Soccer For Dummies
978-0-471-77381-8

Fitness For Dummies, 3rd Edition
978-0-7645-7851-9

Football For Dummies, 3rd Edition
978-0-470-12536-6

GarageBand For Dummies
978-0-7645-7323-1

Golf For Dummies, 3rd Edition
978-0-471-76871-5

Guitar For Dummies, 2nd Edition
978-0-7645-9904-0

Home Recording For Musicians For Dummies, 2nd Edition
978-0-7645-8884-6

iPod & iTunes For Dummies, 5th Edition
978-0-470-17474-6

Music Theory For Dummies
978-0-7645-7838-0

Stretching For Dummies
978-0-470-06741-3

Get smart @ dummies.com®

- **Find a full list of Dummies titles**
- **Look into loads of FREE on-site articles**
- **Sign up for FREE eTips e-mailed to you weekly**
- **See what other products carry the Dummies name**
- **Shop directly from the Dummies bookstore**
- **Enter to win new prizes every month!**

* **Separate Canadian edition also available**
† **Separate U.K. edition also available**

Available wherever books are sold. For more information or to order direct: U.S. customers visit www.dummies.com or call 1-877-762-2974.
U.K. customers visit www.wileyeurope.com or call (0) 1243 843291. Canadian customers visit www.wiley.ca or call 1-800-567-4797.

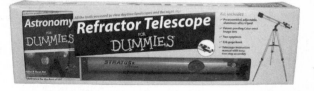